Odor Quality
and Chemical Structure

Odor Quality
and Chemical Structure

Howard R. Moskowitz, EDITOR
Weston Group

Craig B. Warren, EDITOR
International Flavors and Fragrances, Inc.

Based on a symposium

sponsored by the Division of

Agricultural and Food Chemistry

at the 178th Meeting of the

American Chemical Society,

Washington, D.C.,

September 13, 1979.

ACS SYMPOSIUM SERIES 148

AMERICAN CHEMICAL SOCIETY
WASHINGTON, D. C.   1981

Library of Congress CIP Data

Odor quality and chemical structure.
  (ACS symposium series; 148 ISSN 0097-6156)

  Includes bibliographies and index.

  1. Smell—Congresses. 2. Biochemorphology—Congresses. 3. Odors—Congresses.
  I. Moskowitz, Howard R. II. Warren, Craig B.,
1939-   . III. American Chemical Society. Division
of Agricultural and Food Chemistry. IV. American
Chemical Society. V. Series: American Chemical Society. ACS symposium series; 148.

QP458.036   612'.86   80-28633
  ISBN 0-8412-0607-4   ASCMC8   148   1–243   1981

ACS Symposium Series

M. Joan Comstock, *Series Editor*

FOREWORD

The ACS SYMPOSIUM SERIES was founded in 1974 to provide
a medium for publishing symposia quickly in book form. The
format of the Series parallels that of the continuing ADVANCES
IN CHEMISTRY SERIES except that in order to save time the
papers are not typeset but are reproduced as they are sub-
mitted by the authors in camera-ready form. Papers are reviewed under the supervision of the Editors with the assistance
of the Series Advisory Board and are selected to maintain the
integrity of the symposia; however, verbatim reproductions of
previously published papers are not accepted. Both reviews
and reports of research are acceptable since symposia may
embrace both types of presentation.

CONTENTS

PREFACE

O ver the past two decades scientists in many disciplines have become increasingly interested in mechanisms of smell. Researchers from physics, chemistry, biology, psychophysics, and animal behavior have focused their attention on the relation between behavior and chemical structure, each using the techniques of his or her discipline. We still lack an understanding of why chemicals smell the way they do. However, with refined methods of physical and sensory measurement, researchers are beginning to ask the proper questions.

This book presents contributions from a diverse group of researchers interested in the relation between chemical structure and both odor quality and odor intensity. As such, it presents one of the first volumes devoted solely to research in structure–activity relationships, and is a key resource for serious investigators and other interested individuals.

The reader perusing this book, or the researcher using the information for hypothesis building, will notice the variety of interests and focal points represented. Scientists have approached the structure–activity problem from numerous directions. Chapters in this book range from evaluating the contributions of specific characteristics of individual chemicals, to the analysis of different, naturally occurring chemicals, to the development of models for human reactions to odor mixtures. These studies presented in one volume should provide a good launching ground for future research in olfactory science.

HOWARD R. MOSKOWITZ
Weston Group, Inc.
60 Wilton Road
Westport, CT 06880

CRAIG B. WARREN
International Flavors
  and Fragrances, Inc.
1515 Highway 36
Union Beach, NJ 07735

October 13, 1980.

Characterization of Odor Quality Utilizing Multidimensional Scaling Techniques

SUSAN S. SCHIFFMAN

Department of Psychiatry, Duke Medical School, Durham, NC 27710

Research in olfaction has been impeded by a lack of knowl-
ege concerning the physicochemical properties of molecules which
lead to specific olfactory qualities. A diverse range of theories exists which have related quality with physicochemical properties. Factors such as molecular size and shape (1,2), low
energy molecular vibrations (3), molecular cross-section and de-
sorption from a lipid-water interface into water (4), proton,
electron, and apolar factors (5,6), profile functional groups
(7,8), gas chromatographic factors (9), and interactions of the
weak chemical type (10) have all been implicated as variables
related to olfactory quality. Although research investigating
each of these factors has deepened our knowledge of the relationships between odor quality and relevant physicochemical parameters, a strictly predictive model has yet to be achieved.

In the absence of the knowledge of the organizing principles
underlying quality, a technique called "multidimensional scaling"
has proven to be a useful means for studying the organization of
psychophysical and neural data in olfaction. Multidimensional
scaling (MDS) is a mathematical technique which can systematize
data in areas where organizing concepts and underlying dimensions
are not well developed. MDS can represent the similarities of
objects spatially as in a map by utilizing a set of numbers which
expresses all or most combinations of pairs of similarities within a group of objects. Objects judged experimentally similar to
one another are arranged in a resultant spatial map by multidimensional scaling procedures at points close to each other. Objects
judged to be dissimilar are represented at points distant from
one another.

Multidimensional scaling techniques have been successfully
applied to data in color vision. Multidimensional scaling of
both psychophysical data on similarities between colors (11) as
well as spectral absorption data for single cones in the goldfish
retina (12) have produced a color circle. Multidimensional scaling techniques (MDS) have also been helpful in understanding the
full range of the gustatory realm (13-18). Results from such

0097-6156/81/0148-0001$05.25/0

studies suggest that the taste realm extends beyond the traditional sweet, sour, salty, bitter range and is best characterized as
continuous rather than subdivided into four specific groups.

In this paper two sets of psychophysical olfactory data to
which multidimensional scaling techniques were applied are described. In the first study (19,20) (which is based on data from
Wright and Michels (21)) 50 olfactory stimuli, 5 of which were
duplications, were compared with 9 odorant standards, which ranged
widely in quality. The 50 odorants were correlated across the
standards with the assumption that odorants which are highly correlated should have similar smell quality. This 50 x 50 correlation matrix was analyzed by the Guttman-Lingoes' general nonmetric multidimensional scaling technique (22,23). Figure 1 illustrates the two-dimensional space achieved by the Guttman-Lingoes'
method for Wright and Michels' psychophysical olfactory data. The
olfactory stimuli fall roughly into two groups with the larger,
more pleasant subset on the left and an affectively less pleasant
group on the right. Stimuli located near one another in this
space are expected to have more similar olfactory quality than
stimuli located distant from one another. That is, benzeldahyde
and vanillin would be expected to smell more similar to one another than benzaldehyde and pyridine.

It should be noted here that multidimensional scaling procedures attempt to achieve minimum dimensionality. Because of
this feature, the case just described is problematic because there
are only two major clusters. Nonmetric multidimensional scaling
procedures will tend to drive the groups apart and flatten them
out, causing internal relationships within a single cluster to be
lost. For this reason, the two clusters were reanalyzed individually so that any internal relationships which might have been lost
in the arrangement in Figure 1 can be regained in a reanalysis.
The reanalysis of the affectively more pleasant group of stimuli
is shown in Figure 2a, while the reanalysis of the affectively
unpleasant stimuli is shown in Figure 2b.

The spaces were examined with regard to the olfactory qualities traditionally associated with these stimuli utilizing Moncrieff (24) and Merck Index (25) as references for quality descriptions (see Figures 3a and 3b). An examination of the spaces
with regard to traditional qualities indicates that there are no
distinct classes as proposed by many early classifiers of odor
quality. Rather, there appear to be gradual qualitative shifts
in these spaces from one side to the other. For example, in Figure 3a, which corresponds to the stimuli in Figure 2a, qualitative
changes appear to be from a fruity or flowery smell on the right
to a more spiritous or resinous smell on the left. From top to
bottom the quality seems to increase in sharpness or spiciness.
In Figure 3b it is more difficult to find trends because of the
nebulous verbal descriptions given to unpleasant odors. In general, these two figures, 3a and 3b, point out the difficulties encountered in trying to organize olfactory dimensions by means of

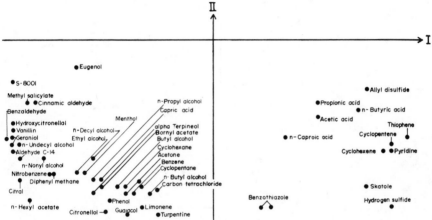

Figure 1.   *Two-dimensional solution, achieved by Guttman–Lingoes' method (22, 23) for Wright and Michels' psychophysical olfactory data for 50 stimuli.*

Substances found by Wright and Michels to be highly correlated are located proximate to one another in this space and are expected to have similar olfactory quality. The more pleasant stimuli are located in the subset on the left, while the more unpleasant stimuli are located in the subset on the right. See Refs. 19 and 20.

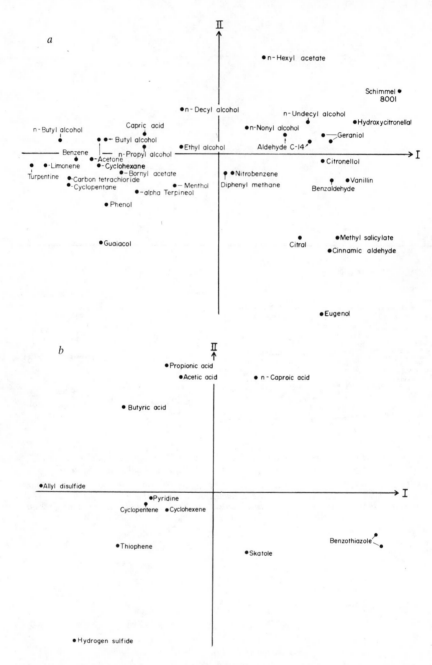

Figure 2.   *Two-dimensional arrangement achieved by reanalysis by Guttman–Lingoes' method (22, 23) of (a) the left-hand, more pleasant cluster in Figure 1 and (b) the right-hand, more unpleasant cluster in Figure 1 (19, 20)*

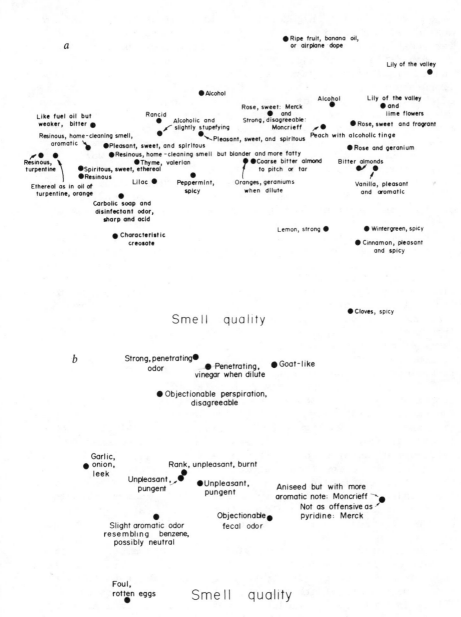

Figure 3. The olfactory qualities traditionally associated with (a) the stimuli in Figure 2a and (b) the stimuli in Figure 2b.

For example, the descriptor "cloves spicy" in the lower right-hand corner of Figure 3a pertains to eugenol, falling in the lower right-hand corner of Figure 2a (19, 20). It can be seen that descriptors for unpleasant smelling stimuli tend to be vague (19, 20).

adjective ratings. In general, people can't articulate olfactory
quality with precision. In addition, there are individual differences in perception as well as in the use of the same words
to mean different things. Application of multidimensional scaling to similarity judgments does not require any a priori assumptions about the dimensions and therefore circumvents the problem
of characterizing olfactory stimuli with adjectives alone. The
use of quantitative experimental measures based on nonverbal
similarity judgments for input to multidimensional scaling procedures is a far more effective means of ordering stimuli to
examine physicochemical dimensions than one based on words (ver-
bal/adjective descriptors).

     The molecular formulae associated with the stimuli in Fig-
ures 2a and 2b are shown in Figures 4a and 4b, respectively. It
can be seen that there are some trends in shape and size with
olfactory quality, but these do not confirm Amoore's specific
shapes (1,2) and thus suggest that stereochemical properties do
not provide the whole answer for predicting olfactory quality.
Several interesting relationships can be seen in these spaces in
Figures 4a and b. In Figure 4a, benzene, cyclopentane, and cyclohexane group together. In Figure 4b, pyridine, cyclopentene,
and cyclohexene group together. The relationship among the three
compounds in Figure 4a is maintained when nitrogen is substituted
into the benzene ring, and when double bonds are added in the
cases of cyclopentane and cyclohexane to yield cyclopentene and
cyclohexene. These changes radically alter olfactory quality
from pleasant to unpleasant.

     The spatial arrangements were examined with regard to func-
tional groups on the odorant molecule. Figure 5a corresponds to
the spatial arrangement in 2a; Figure 5b corresponds to the spatial arrangement in Figure 2b. It can be seen that the aldehydes,
esters, alcohols, ethers, halogens, phenols, and ketones fall
into more pleasant space in Figure 5a. The lightweight carboxylic acids, nitrogens (not associated with oxygen), and sulfurs
fall into less pleasant space in Figure 5b. Thus, although there
are trends in the relationship of functional group to olfactory
quality, functional group alone, like stereochemical properties,
does not provide the entire answer for predicting olfactory quality.

     Next the distribution of molecular weights among the stimu-
li were examined, as shown in Figures 6a and 6b. It can be seen
that the more flowery, fruity odors on the right tend to have
higher molecular weights than the more spiritous odors on the
left. In addition, the molecular weights in the unpleasant space
in Figure 6b have a tendency to be lower than those in Figure 6a.

     The relationship of other physicochemical properties to
these spaces was examined as well. All of the stimuli were ether
soluble, suggesting that fat (ether) solubility may be a necessary requirement for olfactory stimulation to occur. No specific
trends were found for the number of double bonds, dipole moments,

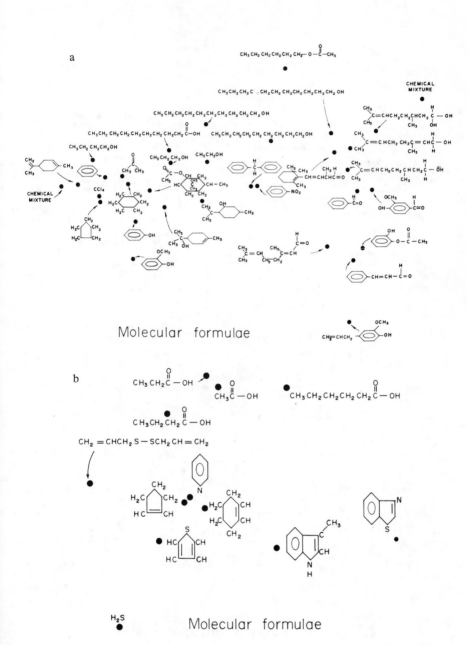

Figure 4.   The molecular formulae associated with the stimuli in (a) Figure 2a and (b) Figure 2b (19, 20)

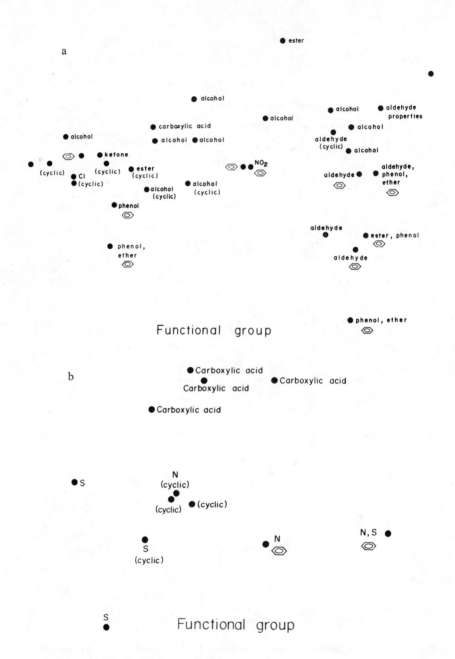

Figure 5.   Functional groups associated with the stimuli in (a) Figure 2a and (b)
Figure 2b (19)

Figure 6.   Distribution of molecular weights for the stimuli in (a) Figure 2a and (b) Figure 2b (19)

water solubilities, or freezing points with olfactory quality.
A relationship was found for Raman spectra, however. Examination
of Raman spectra from 100 cm^{-1} to 1000 cm^{-1} was done to determine if molecules with similar vibrational frequencies have
similar odor quality as suggested by Wright (3). For the stimuli
here, it was found that vibrational frequencies in this range
were highly predictive of the "goodness" or "badness" of the odor
but they were not helpful in further differentiations of the
quality.

The discussion above illustrates that no single physicochemical property is useful on an individual basis in predicting olfactory quality. However, the physicochemical properties discussed above are predictive to some degree in the aggregate when
they are weighted mathematically by a method developed by Schiffman et al. (26). By weighting the physicochemical parameters
shown in Table I, this method was used in an attempt to regener-
ate the space in Figure 1. The correlation between the spatial
arrangement in Figure 7, that is, the theoretical distances
achieved by weighting physicochemical variables, and the original distances shown in Figure 1, which is based on psychophysical
measures, is .76. It can be seen that the variables utilized
here do not produce a perfect regeneration, and therefore some
of the variables necessary to predict olfactory quality must
necessarily be missing from the list in Table I.

Thus, this methodology can be useful in discovering physicochemical variables relevant to olfactory quality in that it
strictly relates quantitative psychophysical measures with quantitative psychophysical chemical measures.

Study 2

In a second experiment (27), 19 odorants were arranged in a
two-dimensional space by ALSCAL (28), another nonmetric multidimensional scaling procedure which can utilize similarity judgments for deriving spaces to map psychological odor quality. The
spatial arrangement for this set of stimuli is shown in Figure 8.

After all the similarity judgments were obtained, each of
the stimuli was rated on a series of adjective scales. It was
found that some of the scales could be related to the space by
regression techniques, and this is illustrated by the vectors
which extend through the space corresponding to the adjective
scales burning, sharp, good, fragrant, putrid, and foul. The
projections of the stimuli on the vectors in Figure 8 are highly
correlated with the mean adjective ratings for subjects on these
scales (see small numbers in parentheses).

A predictive relationship of low energy molecular vibrations
to olfactory quality utilizing a similar range of Raman spectra
as in the previous example was found for this set of stimuli.
The range was divided into 12 intervals of 75 cm^{-1} each. When
the mean intensities for all 12 intervals were weighted mathe-

Figure 7.   Two-dimensional space regenerated from weighting the physicochemical variables shown in Table I in an attempt to reproduce the psychophysical space in Figure 1 (19, 20).

Table I

Weights which were applied to standard scores for physicochemical
variables to achieve the regenerated space in Fig. 7.   Means and
variances for these variables are also given.   Functional groups
are coded according to their number in a molecule; thus, benzaldehyde is coded "1" and the mean number of aldehyde groups for
all the molecules in Fig. 7 is 0.10.   Cyclic compounds are coded
           "1" while noncyclic compounds are coded "0."

Physicochemical variable	Mean	Variance	Weight
Molecular weight	116.57	1788.64	6.24
Number of double bonds	0.74	0.55	0.51
Phenol	0.13	0.11	2.33
Aldehyde	0.10	0.09	3.21
Ester	0.05	0.05	0.24
Alcohol	0.26	0.19	2.54
Carboxylic acid	0.13	0.11	5.50
Sulfur	0.08	0.07	3.44
Nitrogen	0.08	0.07	3.15
Benzene	0.33	0.27	-0.14
Halogen	0.03	0.02	-0.34
Ketone	0.03	0.02	-0.19
Cyclic	0.31	0.21	4.56
Mean Raman intensity			
Below 175 cm^{-1}	0.51	3.14	0.01
176-250 cm^{-1}	2.36	9.30	3.57
251-325 cm^{-1}	1.65	7.10	-0.75
326-400 cm^{-1}	1.56	5.74	3.81
401-475 cm^{-1}	2.10	7.23	1.65
476-550 cm^{-1}	1.54	5.22	-3.63
551-625 cm^{-1}	2.07	7.09	-0.69
626-700 cm^{-1}	1.07	5.14	-1.16
701-775 cm^{-1}	2.36	11.01	0.07
776-850 cm^{-1}	4.36	13.84	3.04
851-925 cm^{-1}	3.44	15.77	0.24
926-1000 cm^{-1}	2.06	8.29	0.36

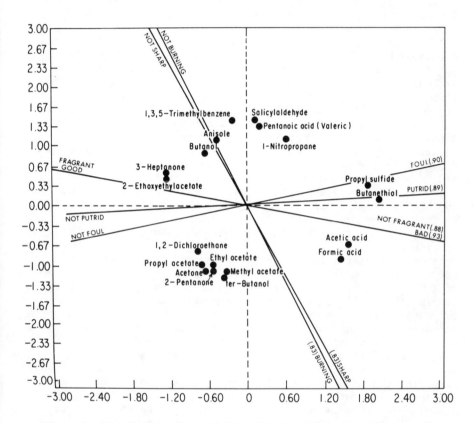

*Figure 8.   Two-dimensional space achieved from experimental measures of simi-
larity among 19 stimuli utilizing ALSCAL (28).*

*Stimuli located near one another are more similar in odor quality. Adjectives were pro-
jected through the multidimensional space by regression techniques. The numbers in
parentheses reflect the correlations between the mean adjective ratings for each of the
stimuli on a semantic differential scale and the projection of the stimuli on the adjective
dimensions (see Ref. 27).*

matically by the same procedure as referred to in the first
study, a correlation of .69 was found between the psychological
distances in Figure 8 and the distances derived from weighted
spectra shown in Figure 9. Thus Figure 9 illustrates the arrangements of stimuli in a space regenerated from weighting Raman
frequencies in an attempt to reproduce the psychological space in
Figure 8. The weights that were applied to the standard scores
for mean Raman intensities to achieve this regenerated space are
given in Table II.

Weights were also applied to standard scores for parameters
developed by Laffort (c.f. 5 and 6) to achieve the regenerated
space in Figure 10. Acetic, formic, and pentanoic acids were
excluded in the calculations because data were incomplete for
these stimuli. The weights utilized to achieve the space in
Figure 10 are given in Table III.

The correlation between the space regenerated from weighting
Raman intensities with the space in Figure 8 is .69. The correlation utilizing the Laffort parameters between the space in
Figure 10 and that in Figure 8 is .40. It can be seen from this
and the previous study that at present we still do not have a
thorough understanding of the physicochemical variables required
to totally predict olfactory quality for stimuli which include
a wide range of odorants.

Multidimensional scaling has been applied to a wide range
of problems in the chemical senses (13-20, 27, 29-38). The direction of research in the author's laboratory is presently
focused in three directions to most effectively exploit the power
of MDS. First, spatial arrangements are being limited to narrow
ranges of stimuli, such as selected pyridyl ketones or substituted pyrazines. Second, intensity dimensions are being introduced to determine qualitative changes with concentration. Third,
the physicochemical parameters are being expanded to parameters
dealing with biological interactions with membranes. Both by
narrowing our scope in the type of spatial arrangements used and
expanding the physicochemical parameters used for prediction,
the methodology of multidimensional scaling may ultimately be
useful in helping us to better understand the relationship between olfactory quality and physicochemical dimensions.

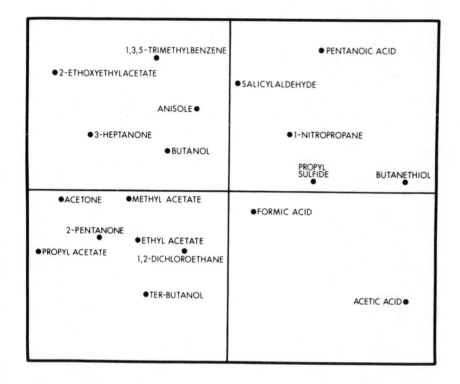

Figure 9.   Two-dimensional space regenerated from weighting Raman frequencies shown in Table II in an attempt to reproduce the psychophysical space in Figure 8 (27).

Table II

Weights that were applied to the standard scores for mean Raman
intensities to achieve the regenerated space in Figure 9 in
Experiment 2

Raman range	Weight
Below 175 cm^{-1}	1.31
176-250 cm^{-1}	6.33
251-325 cm^{-1}	2.49
326-400 cm^{-1}	2.58
401-475 cm^{-1}	6.86
476-550 cm^{-1}	2.28
551-625 cm^{-1}	2.28
626-700 cm^{-1}	1.71
701-775 cm^{-1}	1.89
776-850 cm^{-1}	-1.13
851-925 cm^{-1}	3.67
926-1000 cm^{-1}	3.19

Table III

Weights that were applied to the standard scores for Laffort's
parameters in Experiment 2 to achieve the regenerated space in
Figure 10. Acetic, formic, and pentanoic acids were excluded in
the calculations because complete data were unavailable for these
stimuli

	Weight
Alpha	9.18
(an apolar factor which is proportional to molvolume; relates to van der Waals forces and perhaps surface area of the molecule)	
Rho	4.37
(a proton receptor factor which is relatively high for nitriles and oxygenated and nitro compounds)	
Epsilon	12.46
(an electron factor which is relatively high in cyclic compounds, compounds with double and triple bonds and containing divalent sulfur, bromides, and iodides)	
Pi	0.96
(a proton donor factor which is high in alcohols, two chlorides, and probably primary amines)	

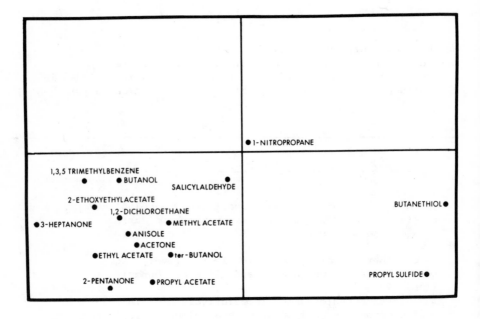

Figure 10. Two-dimensional space regenerated from weighting Laffort's parameters shown in Table III in an attempt to reproduce the psychophysical space in Figure 8. Three acids, acidic, formic, and pentanoic, were not included because Laffort's parameters were not known for these stimuli (27).

Literature Cited

1. Amoore, J. E.   The stereochemical specificities of human olfactory receptors.   Perfum. essent. Oil Rec., 1952, 43, 321-323, 330.
2. Amoore, J. E.; Venstrom, D.   Correlations between stereochemical assessments and organoleptic analysis of odorous compounds.   In T. Hayashi, Ed., "Olfaction and Taste II," Pergamon Press: Oxford, 1967; pp. 3-17.
3. Wright, R. H.   Odour and molecular vibration. I. Quantum and thermodynamic considerations.   J. Appl. Chem., 1954, 4, 611-615.
4. Davies, J. T.   A theory of the quality of odours.   J. Theoret. Biol., 1965, 8, 1-7.
5. Dravnieks, A.   Contribution of molecular properties of odorants to the hedonic value of their odors.   Paper presented at the 7th Symposium of the Sense of Smell, Cannes, France, 1972.
6. Dravnieks, A.; Laffort, P.   Physico-chemical basis of quantitative and qualitative odor discrimination in humans.   In D. Schneider, Ed., "Olfaction and Taste IV," Wissenshaftliche Verlagsgesellschaft MBH: Stuttgart, 1972; pp. 142-148.
7. Beets, M. G. J.   Molecular structure and organoleptic quality.   Soc. Chem. Ind. Monograph, 1957, No. 1, London, p. 54.
8. Beets, M. G. J.   Olfactory response and molecular structure.   In L. M. Beidler, Ed., "Handbook of Sensory Physiology, Vol. IV, Chemical Senses I, Olfaction," Springer-Verlag: Berlin, 1971; pp. 257-321.
9. Mozell, M. M.   The spatiotemporal analysis of odorants at the level of the olfactory receptor sheet.   J. Gen. Physiol., 1966, 56, 25-41.
10. Kafka, W. A.   A formalism on selective molecular interactions.   In L. Jaenicke, Ed., "Biochemistry of Sensory Functions," Springer-Verlag: Berlin, 1974; pp. 275-278.
11. Shepard, R. N.   The analysis of proximities: Multidimensional scaling with an unknown distance function. II.   Psychometrika, 1962, 27, 219-246.
12. Schiffman, H.; Falkenberg, P.   The organization of stimuli and neurons.   Physiol. Behav., 1968, 3, 197-201.
13. Schiffman, S. S.; Dackis, C.   Multidimensional scaling of musks.   Physiol. Behav., 1976, 17, 823-829.
14. Schiffman, S. S.; Erickson, R. P.   A psychophysical model for gustatory quality.   Physiol. Behav., 1971, 7, 617-633.
15. Schiffman, S. S.; Moroch, K.; Dunbar, J.   Taste of acetylated amino acids.   Chem. Senses Flav., 1975, 1, 387-401.
16. Erickson, R. P.; Schiffman, S. S.   The chemical senses: A systematic approach.   In M. S. Gazzaniga & C. Blakemore, Eds., "Handbook of Psychobiology," Academic Press: New York, 1975; pp. 393-426.
17. Schiffman, S. S.   Contribution of the anion to the taste

quality of sodium salts. In M. R. Kare, M. J. Fregley, &
R. A. Bernard, Eds., "Biological and Behavioral Aspects of
NaCl Intake," Nutrition Foundation Monograph Series. Aca-
demic Press: New York, 1980; pp. 99-111.

18. Schiffman, S. S.; McElroy, A. E.; Erickson, R. P. The range
of taste quality of sodium salts. Physiol. Behav., 1980,
24, 217-224.

19. Schiffman, S. S. Contributions to the physicochemical dimensions of odor: A psychophysical approach. Ann. N.Y.
Acad. Sci., 1974, 237, 164-183.

20. Schiffman, S. S. Physicochemical correlates of olfactory
quality. Science, 1974, 185, 112-117.

21. Wright, R. H.; Michels, K. M. Evaluation of far infrared
relations to odors by a standards similarity method. Ann.
N.Y. Acad. Sci., 1964, 116, 535-551.

22. Guttman, L. A general nonmetric technique for finding the
smallest coordinate space for a configuration of points.
Psychometrika, 1968, 33, 469-506.

23. Lingoes, J. C. An IBM 7090 program for Guttman-Lingoes
smallest space analysis-I. Behav. Sci., 1965, 10, 183-184.

24. Moncrieff, R. "The Chemical Senses," CRC Press: Cleveland,
1967.

25. Stecher, P. E.; Windholz, M.; Leahy, D. S.; Boulton, D. M.;
Eaton, L. G. "The Merck Index," Merck and Co.: Rahway,
N.J., 1968.

26. Schiffman, H.; Conger, A.; Schiffman, S. S., unpublished
manuscript; based on Conger, A.; Schiffman, H.; Schiffman,
S. S., a paper presented at the annual meeting of the
Society of Multivariate Experimental Psychology, Educational
Testing Service, Princeton, N.J., 1973.

27. Schiffman, S.; Robinson, D. E.; Erickson, R. P. Multidimensional scaling of odorants: Examination of psychological
and physicochemical dimensions. Chem. Senses Flav., 1977,
2, 375-390.

28. Takane, Y.; Young, F. W.; de Leeuw, J. Nonmetric individual
differences multidimensional scaling: An alternating least
squares method with optimum scaling features. Psychometrika,
1977, 42, 6-67.

29. Schiffman, S. S. Multidimensional scaling: A useful tool
to measure flavor. Cereal Foods World, February, 1976,
64-68.

30. Schiffman, S. S. Food recognition by the elderly. J. Geron.
1977, 32, 586-592.

31. Schiffman, S. S. Changes in taste and smell with age: Psy-
chophysical aspects. In J. M. Ordy & K. R. Brizzee, Eds.,
"Sensory Systems and Communication in the Elderly," (Vol. 6).
Raven Press: New York, 1979; pp. 227-246.

32. Schiffman, S. S.; Dackis, C. Taste of nutrients: Amino
acids, vitamins, and fatty acids. Percept. Psychophys.,
1975, 17, 140-146.

33. Schiffman, S. S.; Engelhard, H. H.   Taste of dipeptides.
     Physiol. Behav., 1976, 17, 523-535.
34. Schiffman, S. S.; Musante, G.; Conger, J.   Application of
     multidimensional scaling to ratings of foods for obese and
     normal weight individuals. Physiol. Behav., 1978, 21,
     417-422.
35. Schiffman, S. S.; Nash, M. L.; Dackis, C.   Reduced olfactory
     discrimination in patients on chronic hemodialysis. Physiol.
     Behav., 1978, 21, 239-242.
36. Schiffman, S. S.; Pasternak, M.   Decreased discrimination
     of food odors in the elderly. J. Geron., 1979, 34, 73-79.
37. Schiffman, S. S.; Reilly, D. A.; Clark, T. B.   Qualitative
     differences among sweeteners. Physiol. Behav., 1979, 23,
     1-9.
38. Schiffman, S. S.; Reynolds, M. L.; Young, F. W.   "Handbook
     of Multidimensional Scaling: Theory and Applications,"
     Academic Press: New York, in press.
39. This paper was supported in part by a grant to the author
     NIA AE 00443.

RECEIVED October 13, 1980.

Psychophysical Scaling and Optimization of Odor Mixtures

HOWARD R. MOSKOWITZ

Weston Group Inc., 60 Wilton Road, Westport, CT 06880

This paper concerns the study of odor mixtures, and their relation to the underlying odor components. Traditionally, chemists, physiologists and psychologists try to relate the quality of an odorant to its chemical structure and to its molecular properties. This paper presents an alternative method, which transcends that stage. It creates mixtures with known constituents, and determines the mixture qualities from the qualities of the components. Thus, this paper presents another direction in the search for the relation between odors and qualities. It espouses a pragmatic approach. Not knowing how molecular structures correlate with odor, it builds in known underlying qualities by mixing together simple chemicals, whose odors by themselves are well defined and can be quantified. The mixture odor quality becomes analogous to the quality of the molecule. The simple odor quality parallels the contributions of the components of a single molecule.

Previous studies of odor mixture have often reported rules for the addition of odor intensities, which conform to a vector model, at least in binary mixtures ($\underline{1},\underline{2},\underline{3},\underline{4}$). Higher order mixtures may or may not generate a total odor intensity which conforms to a vector model (Mixture $^2 = A^2 + B^2 + 2AB \cos\alpha$; A = odor intensity of component A, B = odor intensity of component B, Mixture = odor intensity of the mixture, $\cos\alpha$ = cosine of the angle separating these vectors). Laffort and Dravnieks suggested another ("U") model of additivity which seems more tractable ($\underline{5}$).

The quality and hedonics of a mixture seem less amenable to empirical investigation. Dravnieks et al in this symposium present an elegant approach which relates the complexity of description of a mixture to the complexity of description of the mixture components, evaluated separately. Moskowitz et al ($\underline{6}$) attempted to relate the quality of components of binary mixtures to separate physical intensities, and Moskowitz ($\underline{7}$) attempted with some success to relate the quality of components in binary mixtures to component attribute intensities.

0097-6156/81/0148-0023$08.25/0
© 1981 American Chemical Society

The approach of this study comprises these stages:

● Development of binary mixtures of odors, with either similar or dissimilar odors, and evaluations of the odorants at different levels alone and in mixture.

● Presentation of the odors, in unmixed and mixed form, by means of an air dilution olfactometer, which maintains constant stimulus concentration over long periods of time, independent of the stimuli being smelled or not. The Dravnieks mixture olfactometer provides this capability, and has been previously described (6).

● Scaling of the simple odors and their binary mixtures on a variety of characteristics, including measures of overall odor strength, odor liking/disliking, odor mixture complexity, and 12 additional descriptor characteristics appropriate for the particular odorants studied.

The analytic portion of the study followed this sequence:

● Obtain and average the ratings from the panelists

● Develop a data base, showing odor intensity levels and average magnitude estimate ratings of odor attributes from the panel. (8) The magnitude estimation method is an accepted, very sensitive method which has been previously used to provide the reliable data on the quantitative relations between concentration and perception.

● Develop linear equations relating attribute perception levels and odor concentrations. The equations appear schematically as:

Attribute Intensity = $k_0 + k_1$(Odorant A) + K_2(Odorant B)

Goodness of fit of the equations to the data was indexed by the multiple correlation (R), whose square x 100% gives the percentage of the variability in the ratings accounted for by the equation. (e.g., an R of 0.8 means that 0.8^2 x 100% or 64% of the variability can be accounted for by variations in the levels of the two components).

● Develop non-linear equations (i.e., parabolic equations) to relate overall liking/disliking of odor to the concentrations of the components. The equation is:

Liking = $k_0 + k_1$(Odorant A) + k_2(Odorant A)$^2 + k_3$(Odorant B) + k_4(Odorant B)$^2 + k_5$(Odorant A)(Odorant B)

Goodness-of-fit of the equations was again indexed by the multiple correlation.

• Optimize overall acceptability by maximization of the non-linear liking equation, using standard statistical methods. The optimum combination of odorant levels for odorants A and B was determined, subject to specific constraints:

- The odorant concentrations remained within the 0-64 relative unit ranges tested in the actual experiment. The sensory attributes could act as constraints. For instance, one goal was determination of the optimum acceptability level, with the perception of overall odor intensity lying between prescribed limits of intensity.

• Optimize the closeness of a predicted quality profile to a desired quality profile specified by the experimenter. In concrete terms, the experimenter specified a sensory profile to be achieved (goal profile). The optimum here represents that specific combination of components, within the tested limits, which generates a sensory profile as close as possible to the predesignated goal profile.

Data Base Development

Tables I, II, III and IV show the data base for the four sets of experiments reported there. Note that in each experiment a group of non-expert panelists evaluated each of the sets of odor mixtures twice, using magnitude estimation scaling. Thus, the tables each present numbers which are averages of approximately 32-36 ratings, depending upon the particular study. Furthermore, note that in Tables I-IV, the panelists profiled each stimulus on a variety of sensory characteristics.

Validity of the Ratings

The first analysis of the ratings concerns their validity. Can panelists actually scale the relative sensory impressions of these odor stimuli by magnitude estimation? Correct scaling of overall odor intensity provides a validating measure of the panelist's sensory capabilities in this complicated study. Since panelists had the opportunity to scale unmixed odorants as well as the odor mixtures, and since the unmixed odorants comprised a graded intensity series (albeit presented at random in the set of 24 stimuli) it becomes a straightforward matter to determine whether panelists could pick out the 4 levels of each unmixed odorant, and scale them in the correct order of concentration. Panelists should do so. Table V shows linear and log-log (viz., power functions) relations between odor concentration in air, and rated overall odor intensity, for each pair of odorants in each study. Linear and power functions fit the data adequately. For power functions, the exponents are less than 1.0, confirming previously reported results in the literature. (2, 3)

Quite often researchers in the aroma and fragrance industries claim that panelists cannot possibly evaluate more than just a few odorants, for

TABLE I

Mixtures of Isoamyl Acetate and Amyl Acetate

| ODORANTS | | RESPONSE RATINGS | | | | | | | | | | | | | | |
ISOAMYL ACE TATE	AMYL ACE TATE	INTEN SITY	OVERALL LIKING	COMPLEX ITY	BAN ANA	SWEET NESS	FRUITI NESS	HEAVI NESS	FLOWERY	AROM ATIC	FRAG RANT	ROTTEN	WINEY	GREEN	HER BAL	FERMEN TED
0	1.0	10.0	11.0	7.4	11.4	8.6	10.5	3.8	0.7	5.4	5.0	0.2	0	0	1.4	0
0	4.0	18.1	16.0	9.4	17.9	14.8	11.8	9.1	2.4	8.9	17.6	1.1	2.4	0.9	1.9	1.4
0	16.0	37.3	20.1	18.1	28.1	19.0	26.4	21.8	7.3	18.6	13.4	1.5	1.1	1.6	5.8	3.4
0	64.0	39.3	10.0	22.6	26.6	15.9	25.0	28.3	3.3	15.3	8.0	2.3	3.9	3.1	5.4	5.0
4.0	1.0	17.1	15.3	9.8	12.7	8.1	13.0	6.7	2.0	6.6	8.1	0.5	1.1	0.5	0.9	0.9
4.0	4.0	21.3	13.1	15.5	21.0	15.1	19.1	19.0	4.8	13.6	14.0	0.9	1.8	0.7	1.9	1.8
4.0	16.0	41.1	14.8	17.8	30.0	18.1	29.1	26.5	5.9	10.9	9.9	1.6	1.6	1.9	8.7	2.5
4.0	64.0	54.5	4.1	22.5	39.5	16.6	34.1	39.9	4.5	13.9	10.2	2.4	1.4	1.8	12.2	5.9
1.0	1.0	8.9	8.6	8.1	12.3	7.2	5.5	2.9	1.8	3.6	8.6	0.5	0.9	0.5	0.5	1.3
1.0	4.0	18.4	10.5	10.4	16.8	12.0	13.0	12.5	2.5	11.3	11.7	0.7	0.5	0.5	3.4	1.1
1.0	16.0	27.5	16.6	13.7	23.6	14.3	19.1	13.7	4.5	16.0	12.8	1.4	0.9	0.9	3.3	4.3
1.0	64.0	43.4	11.8	19.9	33.2	16.0	23.5	31.9	3.6	14.1	10.2	3.4	2.9	2.3	10.3	8.5
1.0	0	5.0	7.0	5.1	8.3	4.3	2.5	0.9	0.3	3.1	4.6	0.5	0	0	1.1	0
4.0	0	10.5	11.5	8.1	11.5	7.3	10.4	2.2	0.7	5.9	6.3	0.6	0.5	0.5	2.3	0.4
16.0	0	35.5	0.7	21.0	25.9	16.1	21.5	20.7	6.3	16.0	13.6	1.8	0.9	0.9	6.4	3.4
64.0	0	52.7	3.7	24.4	36.8	16.8	25.1	35.8	7.3	16.0	6.8	2.3	1.8	1.8	12.6	3.9
16.0	1.0	33.2	18.6	20.9	27.4	14.5	22.0	20.2	2.3	18.6	16.1	1.4	3.0	0.9	3.9	3.2
16.0	4.0	28.4	12.4	12.1	22.0	15.3	11.6	17.8	5.9	11.6	10.4	1.6	1.6	0.9	6.6	5.5
16.0	16.0	35.5	16.1	18.0	31.6	20.7	32.5	28.6	5.4	16.4	14.9	2.5	1.1	1.4	6.6	4.0
16.0	64.0	48.4	11.3	22.9	24.7	19.5	28.6	37.2	5.9	20.9	12.5	3.0	4.3	3.6	6.7	8.6
64.0	1.0	50.5	3.1	20.6	29.3	14.9	24.7	39.8	2.7	20.9	7.3	3.2	3.4	1.6	9.4	6.2
64.0	4.0	50.0	-4.5	23.0	28.3	21.1	27.5	36.0	5.0	21.1	10.9	2.0	1.9	3.0	10.1	3.6
64.0	16.0	54.2	-6.4	23.7	36.7	19.3	29.7	41.0	6.3	23.4	8.9	3.0	1.8	4.0	14.3	4.3
64.0	64.0	58.5	-7.9	25.2	37.2	21.4	30.2	44.2	7.1	24.1	9.4	3.0	1.8	4.2	20.7	6.8

TABLE II

Mixtures of Methyl Salicylate and Ethyl Salicylate

ODORANTS		RESPONSE RATINGS														
METHYL SALICYLATE	ETHYL SALICYLATE	INTENSITY	OVERALL LIKING	COMPLEXITY	CARNATION	FLORAL	GREEN	SPICY	MINTY	SHARPNESS	WINTERGREEN	MEDICINAL	HEAVINESS	FLOWERY	PEPPERMINTY	FRUITINESS
0	1.0	8.6	18.6	20.0	6.8	7.1	4.4	7.3	10.3	5.7	9.2	6.6	7.0	7.8	8.5	6.9
0	4.0	14.1	16.5	20.6	9.9	7.1	6.1	10.2	11.7	8.8	13.1	7.1	8.5	7.1	11.1	6.5
0	16.0	18.8	25.6	24.7	10.4	9.2	7.1	10.9	21.1	8.9	20.6	9.2	8.6	9.1	14.3	12.8
0	64.0	37.1	26.9	30.6	9.2	6.4	11.1	16.7	30.3	19.2	35.0	18.9	16.3	6.5	22.5	11.2
4.0	1.0	15.9	22.9	30.1	9.4	10.8	6.0	8.5	12.1	8.7	15.1	8.6	6.9	9.6	13.3	9.7
4.0	4.0	17.5	26.7	28.8	7.2	9.4	8.3	11.1	18.4	11.2	22.9	9.0	8.8	8.3	13.9	12.2
4.0	16.0	26.4	29.7	29.1	12.5	7.6	7.4	11.5	18.4	13.4	26.8	11.3	9.5	7.5	16.6	10.7
4.0	64.0	41.9	25.8	29.9	8.3	7.3	10.4	16.5	30.6	22.4	36.3	17.2	16.9	6.3	24.3	12.6
1.0	1.0	13.4	22.1	17.5	8.4	7.9	6.3	7.7	12.2	8.6	16.2	6.5	8.5	6.5	8.5	9.2
1.0	4.0	11.4	20.5	19.1	11.5	10.1	6.1	7.9	12.6	9.4	14.8	5.6	7.3	7.6	8.4	8.3
1.0	16.0	25.9	25.1	25.8	8.8	9.1	8.2	9.9	21.2	14.1	22.1	12.2	12.0	9.2	14.5	13.6
1.0	64.0	43.1	23.9	30.7	7.1	6.5	12.4	21.5	33.1	24.3	39.0	19.2	23.6	6.6	22.3	11.5
1.0	0	11.0	21.4	23.4	8.8	8.2	5.4	7.5	10.0	6.9	11.0	6.3	7.2	6.9	7.8	6.9
4.0	0	14.3	23.9	21.7	7.5	5.8	6.4	8.8	14.9	7.2	14.7	7.1	7.5	5.8	10.1	10.0
16.0	0	31.4	25.0	27.2	7.6	7.5	8.2	15.7	31.7	18.9	34.6	14.1	11.8	7.8	18.6	11.9
64.0	0	47.8	28.0	23.4	10.9	6.6	12.4	21.1	33.8	25.0	48.1	16.9	14.9	6.8	30.3	11.2
16.0	1.0	26.3	30.1	26.9	11.7	8.4	7.2	13.9	24.5	22.3	28.2	11.6	11.1	7.2	20.0	11.1
16.0	4.0	19.9	27.3	30.5	11.1	8.9	8.1	13.6	22.3	23.2	26.2	10.8	9.7	9.0	18.5	10.4
16.0	16.0	26.9	27.3	30.9	10.8	9.6	9.0	14.7	21.0	22.8	24.1	11.4	13.4	8.8	17.4	11.5
16.0	64.0	44.7	23.9	30.5	7.5	8.2	12.6	19.0	29.9	25.7	35.8	17.1	18.2	6.7	23.6	13.6
64.0	1.0	45.9	30.8	29.2	8.5	7.4	14.0	21.6	32.6	23.6	45.0	15.1	19.4	6.7	23.5	12.9
64.0	4.0	437	28.0	30.9	8.4	6.5	9.9	18.5	33.6	23.1	41.0	14.3	17.8	6.8	23.9	11.8
64.0	16.0	47.4	25.2	28.5	9.4	6.4	11.6	22.5	34.3	25.3	44.6	15.6	14.9	6.6	25.9	14.2
64.0	64.0	56.7	21.9	31.4	6.7	6.4	9.9	25.3	36.3	28.5	46.7	20.7	20.8	6.5	31.4	10.8

TABLE III

Mixtures of Amyl Acetate and Ethyl Salicylate

| ODORANTS | | ATTRIBUTE RATING | | | | | | | | | | | | | | |
AMYL ACETATE	ETHYL SALICYLATE	INTEN SITY	LIKING	COMPLEX ITY	FRUITI NESS	FRAG RANT	MINTY	BANANA	SWEET NESS	AROMATIC	PEAR	FLOW ERY	HEAVI NESS	SPEAR MINT	WINEY	HER BAL
0	1.0	8.6	11.3	27.1	15.1	0	10.4	0.3	2.4	9.6	0.7	3.9	3.3	5.0	0.4	0.2
0	4.0	21.3	23.0	31.7	28.6	0.7	18.6	2.3	6.8	16.9	3.1	11.7	8.7	19.1	1.5	1.0
0	16.0	31.9	25.6	35.0	35.3	1.8	25.5	3.8	8.2	23.4	4.4	17.9	13.1	24.1	4.5	3.5
0	64.0	42.0	27.9	35.5	42.8	3.7	27.6	4.5	9.8	27.9	5.3	21.2	19.6	32.3	4.8	5.5
4.0	1.0	13.7	12.3	28.1	13.9	0	9.2	3.4	2.8	9.5	3.4	5.0	5.2	9.0	0.8	0.5
4.0	4.0	20.0	23.4	32.8	29.5	0.8	16.7	1.5	6.8	15.8	1.8	9.5	12.2	17.6	3.4	1.3
4.0	16.0	34.8	30.0	34.4	37.3	2.1	25.6	4.0	11.0	25.0	4.9	19.2	16.4	26.2	5.1	2.1
4.0	64.0	44.3	22.7	31.3	43.9	5.3	24.7	3.0	10.0	31.0	4.1	22.7	24.5	30.4	4.1	8.2
1.0	1.0	9.9	9.4	22.5	12.3	0.7	10.2	1.0	3.6	5.5	1.2	5.5	4.2	10.1	0.8	0.4
1.0	4.0	24.5	28.2	26.7	35.4	0.7	24.9	1.9	8.2	23.7	3.5	16.7	11.9	18.9	2.8	1.7
1.0	16.0	36.4	30.3	32.1	46.0	1.3	30.0	4.1	9.7	28.0	5.0	22.4	14.4	26.2	4.3	4.4
1.0	64.0	52.2	21.5	36.3	43.5	7.6	29.1	5.6	9.4	30.4	9.5	21.0	26.9	30.9	6.0	8.7
1.0	0	5.1	4.8	21.8	3.9	0	2.3	1.0	0.8	3.3	2.1	2.9	2.5	0.6	0	0
4.0	0	9.3	9.4	25.1	6.0	0	5.2	5.8	0.6	5.1	3.7	6.2	1.6	5.4	2.1	0
16.0	0	24.5	23.5	41.3	4.9	1.5	11.7	29.3	4.0	22.7	12.9	12.9	7.0	6.2	2.5	0.9
64.0	0	36.0	28.9	38.1	4.5	5.6	16.3	36.0	6.2	26.3	29.8	23.3	13.7	13.7	9.5	2.1
16.0	1.0	26.8	22.8	35.9	13.5	2.4	13.2	16.8	3.8	17.9	8.9	12.0	9.9	11.2	5.9	2.3
16.0	4.0	27.6	24.0	34.4	25.1	0.5	21.1	8.4	8.5	24.7	6.9	15.9	10.1	16.2	3.5	0.7
16.0	16.0	33.7	25.2	33.2	34.1	1.5	25.2	9.2	10.2	29.1	7.5	18.9	16.7	25.3	4.2	2.5
16.0	64.0	51.0	18.5	33.5	48.5	8.9	29.5	5.5	8.9	28.1	10.5	23.1	29.4	26.4	8.0	9.4
64.0	1.0	35.2	26.8	41.1	4.3	1.5	13.8	33.1	3.4	23.5	23.9	14.5	14.3	8.7	6.3	0.8
64.0	4.0	33.5	24.7	35.2	10.4	4.5	21.2	30.9	5.5	25.9	21.5	16.7	17.7	15.2	6.1	3.9
64.0	16.0	38.5	21.2	41.4	24.1	6.7	23.3	20.5	6.8	24.8	13.3	19.5	19.5	16.0	6.3	4.5
64.0	64.0	59.7	19.2	31.7	47.6	6.6	34.1	13.9	10.2	33.5	10.4	24.7	33.8	34.3	8.2	12.7

TABLE IV

Mixtures of Heptyl Acetate and Amyl Acetate

| ODORANTS | | ATTRIBUTE RATINGS | | | | | | | | | | | | | | |
HEPTYL ACETATE	AMYL ACETATE	ODOR INTENSITY	OVERALL LIKING	COMPLEXITY	BANANA	SWEETNESS	FRUITINESS	HEAVINESS	FLOWERY	AROMATIC	FRAGRANT	PEAR	WINEY	ROTTEN	HERBAL	GREEN
0	1.0	12.7	13.2	22.9	16.3	12.7	8.9	6.1	2.9	6.1	6.9	8.2	2.0	1.8	3.0	1.8
0	4.0	17.1	19.8	32.0	28.9	14.6	19.3	8.8	4.3	8.0	8.8	16.0	3.4	3.9	7.0	2.5
0	16.0	37.5	21.4	33.2	41.4	24.1	34.1	26.1	4.5	20.9	15.4	19.7	5.9	3.9	10.9	2.5
0	64.0	52.1	15.2	33.0	47.1	30.0	45.0	43.2	5.2	31.3	20.4	16.4	12.5	8.8	12.1	2.5
4.0	1.0	12.1	17.1	29.0	18.0	9.3	14.1	7.1	1.8	7.3	8.2	7.9	2.0	2.3	4.3	1.8
4.0	4.0	24.7	23.6	30.7	33.0	23.6	27.0	15.2	4.0	20.0	13.4	17.3	4.3	3.0	5.7	2.5
4.0	16.0	41.1	22.3	33.9	40.2	28.5	40.3	27.7	5.5	27.3	23.2	23.8	6.6	2.5	9.6	2.5
4.0	64.0	58.2	15.4	36.6	48.9	30.4	43.6	44.3	6.4	36.3	26.4	19.1	16.6	8.2	10.2	2.5
1.0	1.0	12.2	14.1	20.0	18.9	13.6	15.2	6.1	3.5	8.6	9.5	2.7	2.9	1.8	4.5	1.8
1.0	4.0	22.3	26.6	33.0	37.1	23.4	34.4	11.8	7.7	14.0	17.0	13.0	4.8	3.0	7.3	2.5
1.0	16.0	33.9	21.6	33.9	41.4	26.3	33.9	25.4	5.2	21.8	15.9	16.1	9.6	4.3	8.4	2.5
1.0	64.0	38.9	-.90	25.7	38.5	14.0	28.0	26.0	1.7	23.0	18.3	17.0	14.8	8.7	2.5	0
1.0	0	8.6	20.6	23.9	7.5	11.4	7.2	3.6	3.6	6.1	5.3	4.2	2.8	2.8	3.6	2.8
4.0	0	5.4	12.0	20.9	5.2	4.5	4.5	3.4	1.8	2.3	3.8	2.3	1.8	1.8	2.5	1.8
16.0	0	8.5	14.8	23.9	7.9	4.5	7.2	2.7	1.8	6.6	7.0	2.8	3.2	3.2	2.8	1.8
64.0	0	14.7	13.8	28.9	13.9	8.4	13.5	10.3	1.7	11.8	9.5	6.4	4.5	8.0	3.8	1.9
16.0	1.0	21.6	26.4	36.6	24.6	19.3	22.3	11.1	3.6	13.0	13.4	18.9	5.0	2.1	4.6	2.1
16.0	4.0	24.3	25.4	34.5	23.4	19.3	23.0	13.8	5.4	17.3	14.3	17.7	3.1	1.8	8.4	1.8
16.0	16.0	38.0	26.7	26.8	35.2	37.1	27.0	38.0	28.0	9.0	21.3	19.9	8.2	4.3	12.8	2.5
16.0	64.0	54.8	15.2	34.8	54.1	28.9	45.7	42.3	8.0	34.3	21.6	13.6	14.6	9.6	10.9	2.5
64.0	1.0	17.3	19.6	33.8	14.2	10.4	17.1	10.2	1.1	12.0	9.5	13.3	3.9	2.0	3.0	0.2
64.0	4.0	25.7	18.8	34.3	23.4	22.9	30.6	14.6	5.2	19.3	14.1	18.2	0.9	2.5	10.6	0.7
64.0	16.0	50.2	16.8	40.2	46.3	30.5	43.8	36.4	3.4	33.4	21.1	17.9	8.6	3.4	9.8	0.7
64.0	64.0	66.9	-.70	32.4	55.4	30.9	53.9	49.3	13.1	46.5	28.6	19.8	15.4	13.5	2.6	0

TABLE V

LINEAR AND POWER FUNCTIONS RELATING
SENSORY ODOR INTENSITY AND CONCENTRATION

<u>Mult R</u>

EXPI	Amyl Acetate	Linear	$= 0.37(C) + 18.1$	.77
		Power	$= 10.96(C)^{0.34}$	.96
	Iso Amyl Acetate	Linear	$= 0.69(C) + 11.13$	.92
		Power	$= 5.1(C)^{0.67}$	.98
EXP II	Methyl Salicylate	Linear	$= 0.55(C) + 14.4$	.95
		Power	$= 10.11(C)^{0.37}$	.98
	Ethyl Salicylate	Linear	$= 0.44(C) + 9.76$	.99
		Power	$= 8.46(C)^{0.34}$	.99
EXP III	Ethyl Salicylate	Linear	$= 0.43(C) + 16.9$	.87
		Power	$= 10.26(C)^{0.37}$	.96
	Amyl Acetate	Linear	$= 0.44(C) + 9.2$	.92
		Power	$= 5.13(C)^{0.49}$	.99
EXP IV	Amyl Acetate	LInear	$= 0.58(C) + 17.59$	.92
		Power	$= 12.01(C)^{0.36}$	.98
	Heptyl Acetate	Linear	$= 0.12(C) + 6.7$	.95
		Power	$= 6.41(C)^{0.15}$	.95

these panelists surely adapt and lose their sensitivity to odor stimuli. The present results belie that claim. Panelists evaluated a total of 24 samples, varying extensively in odor intensity from weak to strong, in totally random order. The key to adequate sensitivity may lie in a combination of motivated panelists (who can participate for extended periods of time), and a testing regimen which allows panelists sufficient inter-stimulus time (e.g., 3 minutes or so) to recover their sensitivity. With such a procedure no doubt the enterprising researcher can test far more than 24 stimuli in a session, without substantial changes in panelist sensitivity. The sessions here each lasted about 2 hours, with approximately 4 minutes between samples. This testing regimen promotes sensitivity.

Linear Functions for Attributes
===

Prior to optimization, we first develop a set of linear functions to relate attribute intensities to a linear combination of the two odorants. The general form of the linear function is:

Attribute Intensity = $k_o + k_1 A + k_2 B$

The concentrations of the two odorants are expressed in commensurate terms (in terms of relative amounts in vapor). Thus the coefficients k_1 and k_2 indicate the relative importance and directionality of each component as it affects the intensity of the specific attribute.

Table VI (A-D) shows the coefficients of the four sets of linear equations, one set per experiment. Next to each set of coefficients is the partial correlation which shows how much the specific odorant in the pair contributes to explaining the variability of the attribute ratings. Each equation generates a multiple correlation, as an index of goodness of fit.

Linear equations model some of the attributes quite well, but fail to model other attributes, for at least two possible reasons:

● The data requires a more complicated function to model it, such as a quadratic function (with or without cross terms). Liking/disliking ratings often require a quadratic function.

● The data defy modelling, because the numbers scatter apparently at random. This outcome occurs when panelists have no concept of the meaning of a specific attribute. One panelist may rate a specific stimulus 'high' on that characteristic, whereas another panelist may rate the same stimulus 'low' on the same characteristic. Quite often inappropriate attributes for the specific odor stimuli generate such random appearing functions, with relatively low slopes, and low correlations.

Linear functions are important for modelling odor quality. They provide the researcher with a numerical measure of how odorant concen-

TABLE VI (A)

Amyl Acetate and Isoamyl Acetate
Linear Regression Equations

| | | | | | Partial Correlation | |
	Inter- cept	Isoamyl Ace- tate	Amyl Ace- tate	Mult. Corre- lation	Isoamyl Acetate	Amyl Acetate
Intensity	18.52	.46	.38	.88	.68	.55
Liking	14.70	-.24	-.07	.82	-.79	-.20
Complexity	11.34	.16	.14	.83	.62	.54
Banana	17.54	.21	.20	.77	.56	.52
Sweet	11.60	.10	.09	.68	.50	.45
Fruity	14.12	.17	.20	.73	.46	.56
Heavy	9.82	.39	.33	.91	.69	.58
Flowery	2.90	.04	.03	.55	.45	.30
Aromatic	9.25	.17	.10	.79	.68	.39
Fragrant	10.87	-.02	.00	.18	-.18	.02
Rotten	.86	.02	.03	.87	.58	.64
Winey	1.04	.01	.02	.59	.26	.52
Green	.43	.03	.03	.92	.60	.68
Herbal	1.93	.15	.11	.91	.74	.52
Fermented	1.58	.04	.08	.84	.36	.75

Table VI (B)

Methyl Salicylate and Ethyl Salicylate
Linear Regression Equations

	Inter-cept	Methyl Salicy-late	Ethyl Salicy-late	Mult. Corre-lation	Partial Correlation Methyl Salicy-late	Partial Correlation Ethyl Salicy-late
Intensity	14.68	.44	.35	.96	.74	.59
Liking	23.93	.05	.00	.37	.37	-.01
Complexity	24.02	.06	.09	.63	.33	.52
Carnation	9.57	-.00	-.02	.33	-.06	-.33
Floral	8.54	-.02	-.02	.51	-.41	-.30
Green	6.51	.07	.06	.82	.61	.53
Spicy	9.03	.17	.12	.94	.77	.54
Minty	15.21	.26	.20	.88	.69	.53
Sharp	8.75	.21	.18	.94	.71	.60
Wintergreen	16.92	.39	.24	.92	.78	.46
Medicinal	7.91	.10	.14	.91	.53	.74
Heavy	8.00	.11	.14	.90	.55	.70
Flowery	7.91	-.01	-.01	.47	-.31	-.34
Peppermint	11.41	.21	.15	.92	.74	.53
Fruity	9.74	.03	.03	.54	.39	.36

TABLE VI (C)

Amyl Acetate and Ethyl Salicylate
Linear Regression Equations

	Inter-cept	Amyl Ace-tate	Ethyl Salicy-late	Mult. Corre-lation	Partial Correlation Amyl Ace-tate	Ethyl Salicy-late
Intensity	17.32	.25	.47	.90	.42	.79
Liking	19.5	.07	.04	.28	.23	.15
Complexity	30.05	.12	.03	.56	.54	.15
Fruity	19.42	-.15	.49	.81	-.25	.78
Fragrant	.29	.05	.08	.89	.47	.74
Minty	14.27	.05	.25	.73	.14	.71
Banana	5.10	.38	-.09	.86	.84	-.21
Sweetness	5.08	.00	.08	.65	.-01	.65
Aromatic	14.89	.14	.22	.74	.38	.63
Pear	3.92	.26	-.01	.85	.85	-.05
Flowery	10.18	.11	.18	.77	.37	.66
Heavy	6.40	.14	.30	.94	.38	.85
Spearmint	12.33	-.00	.32	.82	-.03	.82
Winey	2.10	.07	.05	.83	.68	.46
Herbal	.35	.04	.13	.95	.25	.91

TABLE VI (D)

Heptyl Acetate and Amyl Acetate
Linear Regression Equations

	Inter-cept	Heptyl Ace-tate	Amyl Ace-tate	Mult. Corre-lation	Partial Correlation Heptyl Ace-tate	Partial Correlation Amyl Ace-tate
Intensity	15.78	0.14	0.61	0.88	0.18	0.87
Liking	21.48	-0.64	-0.10	0.60	-0.21	-0.58
Complexity	28.14	0.07	0.06	0.46	0.35	0.28
Banana	21.33	0.02	0.48	0.78	0.01	0.78
Sweetness	15.98	0.03	0.21	0.55	0.06	0.55
Fruity	14.79	0.11	0.42	0.77	0.19	0.74
Heavy	9.18	0.11	0.52	0.88	0.16	0.85
Flowery	4.26	0.11	0.05	0.25	0.04	0.25
Aromatic	9.00	0.14	0.38	0.86	0.28	0.81
Fragrant	10.81	0.05	0.21	0.79	0.17	0.76
Pear	11.30	0.03	0.11	0.44	0.12	0.42
Winey	3.15	0.01	0.19	0.95	0.02	0.95
Rotten	1.94	0.03	0.11	0.90	0.22	0.87
Herbal	5.14	-0.11	0.04	0.31	-0.08	0.30
Green	2.37	-0.02	-0.01	0.69	-0.67	-0.17

tration impacts on the perceptions, and they show the relative impact of the two odorants. Furthermore, the linear equation can be used to model overall odor intensity, and to see whether or not the odor mixture intensity smells as strong on the average, as the arithmetic sum of the component odor intensities. (By and large it does not. The mixture odor intensity almost always smells weaker than the arithmetic sum of the odor intensities of the components).

Reversing The Equations - Fitting A Profile

Linear equations conveniently summarize how concentrations of odor components relate to the mixture intensity of specific character-istics. Given the component concentrations, even at intermediate, non-tested levels, one can estimate the profile of perceptions expected from that mixture by using the equations in Table VIA-VID.

Let us turn the problem around, 180 degrees, and reverse the question. Let us specify a profile of sensory perceptions, and estimate what concentrations of the two components which, in concert, produce the goal profile perceptions, or at least come as close as possible to doing so.

By reversing the regression procedure, using the method of multiple objective programming, one can ascertain the specific concentrations of mixture components which come as close as possible to reproducing a desired sensory profile. Of course, in order to get meaningful data, the investigator must make sure that:

● The equations relating sensory characteristics and odorant ingre-dient levels provide at least a reasonably good set of predictors with good multiple correlations (e.g., around 0.80 or so for each equation, although some equations will be better predictors than others).

● The desired levels of the sensory attributes lie within achievable ranges, rather than lying outside of the range spanned by the actual stimuli. One cannot create a combination of odorants which generate unusually high or low levels of specific sensory characteristics, if none of the stimuli generate sensory magnitudes near the high level desired. Furthermore, since we deal with linear equations, rather than with quadratic or other non-linear equations, seeking an unduly high level of a sensory characteristic forces the level of odorant concentrations to 'pin' at the highest allowable or at the lowest allowable concentration.

Table VII shows some hypothetical "desired" sensory profiles for these experiments, as well as the expected sensory profile one could empirically obtain, along with the combination of odorants which come as close as possible to generating that desired profile (as obtained from the multiple objective programming method). To generate these specific profiles, one often must use intermediate levels of each odorant not directly evaluated. Since, however, the researcher has equations in Table VI which relate component concentrations of perceptions of attributes, it becomes a straightforward matter to estimate the likely sensory profile of the mixture.

One can extend the goal profiling method to situations in which one investigates several different odorants (or complex perfumer or flavorist subs) in mixtures, in order to simulate more real world conditions. The technique does not apply solely to two components, but can be generalized in a straightforward manner to mixtures comprising 3,4,5 and even 6 or more components.

Discussion of Profile-Fitting

The foregoing data suggests that it is possible to develop odor mixtures which reproduce a sensory profile if the components possess their specific odor characteristics   Four observations are in order, however.

First, the approach shows that one can engineer a mixture with specific sensory characteristics, by mixing together consitutents which already possess some degree of those characteristics. Rarely can this system accomodate the unique instance of an entirely new odor quality arising from the mixture. The system is synthetic, but not creative.

Second, the system is testable. One can construct the mixtures in order to evaluate their sensory profile. In that respect a mixture system for odor qualities presents the opportunity for further test and validation, which some other methods do not provide.

Third, the approach requires a tradeoff between different desired profile attributes. Sometimes one may specify a combination of attributes impossible to satisfy. The mixing and profile fitting system outlined above will the generate a combination of odorants which achieves certain profile levels, but leaves other attribute levels unsatisfied.

Fourth, the approach bears on the issue of the psychology of odor description and perception. Let us hypothesize the existence of two individuals, each participating in a scaling experiment involving odors. Each individual smells the odorants, scales his or her perceptions of each of the 24 odorants on specific characteristics and then describes the desired odor in terms of the same scales and the same attributes used to profile the actual set of 24 stimuli. Let us suppose that the individuals do not share any language at all. One individual speaks English only, and the other individual speaks only Tagalog (a Philipino dialect). The words in English were translated for the benefit of the Tagalog Speakers, but these two individuals have no other contact. Further assume that each of these two individuals assign an ideal profile based upon some common odor concept (e.g., description of an object) or smell another odor stimulus, and rate this odor stimulus (the 25th) in the same way that they rated the 24

TABLE VII (A)

Profile Fitting to a Predesigned Sensory Profile
Experiment I

| | | | CONCENTRATION OF | |
Attribute	Desired Level	Obtain-able Level	Isoamyl Acetate	Amyl Acetate
(A)				
Intensity	20	20	0	3.92
Banana	40	18.3		
(B)				
Intensity	40	40	0	56.9
Banana	40	28.7		
(C)				
Intensity	60	60	37.9	64.0
Banana	40	37.9		
(D)				
Intensity	20	29.4	0	28.9
Banana	40	23.2		
Fruity	20	20.0		

TABLE VII (B)

Experiment II

Attribute	Desired Level	Obtain-able Level	CONCENTRATION OF Methyl Salicylate	Ethyl Salicy-late
Intensity	25	25.0	23.45	0.00
Floral	40	8.0		
Spicy	25	13.07		
Sharp	40	13.74		

TABLE VII (C)

Experiment III

Attributes	Desired Level	Obtain-able Level	CONCENTRATION OF Amyl Acetate	Heptyl Acetate
(A)				
Intensity	20	20	0	29.71
Banana	20	21.8		
(B)				
Intensity	40	40	1.79	39.23
Banana	40	40		

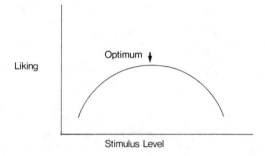

Figure 1. *Relation between the physical concentration or the perceived sensory intensity of an odorant or fragrance (abscissa) and the rated overall liking of the odorant (ordinate) as exemplified by a linear equation and by a quadratic equation.*

The linear equation requires that liking increase with increasing concentration, and not reach a bliss or optimal point in the middle of the concentration range (beyond which further increases in concentration or in sensory intensity only diminish liking). The parabolic, second-order equation allows liking first to increase with intensity, reach a peak level or bliss point, and then diminish with further increases in concentration. The parabolic equation, rather than the linear equation, captures the empirical relation between liking and physical concentration found in these studies. It also reflects the general behavior of liking vs. intensity for other sensory continua as well.

test stimuli. This 25th stimulus is not a member of the set. It could be an ideal profile, or a profile representing some object common to the two cultures, or even an actual odor stimulus presented to the two culturally-different individuals.

Given the set of 24 odorants, the researcher can develop a separate set of equations for each person. The equation relates quality character-istics to a linear combination of the two test odorants, much the same as in Table VI. We expect the equations to differ from one person (the English speaker) to the other (the Tagalog speaker). Each person will have a different conception of what the descriptor terms mean. Furthermore, since the individuals use their descriptive terms differently, we expect them to profile the 25th stimulus differently as well, whether this stimulus be a conceptual ideal, or the profile assigned to an object in a picture (e.g., the smell of a familiar animal, or the actual odor stimulus). Furthermore we do not know what, in fact, each descriptor term of the 15 means either to the U.S. panelist of the Tagalog-Speaking Panel. In effect these represent simply verbal statements whose meanings are left up to the individual.

Despite the differences in profiles and equations, we may well end up with the same set of odor concentrations which in concert produce that ideal profile. Although the individuals differ in their language and their scales, nonetheless it is quite possible that the net combination of ingredients, or the "odorant" recipe could be identical for each person. This test has not been carried out for odor mixtures, but it has been carried out for mixtures of rye flour and sugar, in a study on bread texture (9). Panelist who were experts in the use of the Texture Profiling Method (10), and consumer panelists each evaluated 12 samples of rye bread, profiling the samples on different textural characteristics. After-wards, each group profiled its "ideal" rye bread, on the same character-istics. The linear equations and the ideal profile differed from group to group, but the physical formulation corresponding to the ideal was remarkably similar from group to group.

Odor Acceptability

A centry ago, the German psychologist Wilhelm Wundt (11) specu-lated that as any stimulus increased in sensory intensity, it changed hedonic tone. Beginning at neutral, the stimulus first increased in acceptability, going towards a bliss point, where it maximized. As the stimulus intensity, and thus the sensory intensity further increased, liking diminished from that bliss point, going downwards towards neutrality, and then onto the region of 'dislike'. Figure 1 shows a schematic of the hypothesis.

Wundt's scheme characterizes some odors, but not others. (7,12,13) In many instances odor liking vs concentration does not describe an inverted U or V shaped function. Rather, for the more noxious odorants, liking diminishes almost immediately as the odor intensity increases, going from neutrality (at no odorant level) into disliking.

Wundt's scheme applies to foods, and to complex perfumes as well as to simple chemical stimuli (13). We now wish to assess whether or not Wundt's scheme applies to mixtures, to determine the following:

- The nature of the hedonic function for odor mixtures
- The existence of interactions in mixtures which may modify the concentration of the bliss point
- The bliss point for the odors
- The optimization of liking in odor mixtures, subject to engineering specific sensory characteristics to lie within pre-specified values (e.g., maximize liking, with overall odor intensity lower than a prespecified level).

For some, but not all of the odor mixtures, the relation between overall liking vs concentration can be improved if one uses the non-linear quadratic equation, discussed previously (Equation 2). This non-linear equation allows liking of the odor mixture to increase, peak at an intermediate 'bliss point' of concentrations, and then to drop back down with further increases in concentration.

Table VII presents the non-linear equations, for the four sets of data. Note that the non-linear equations always fit the results better than the linear equations do, in part because the more predictors one can use in the equation the higher will be the multiple correlation coefficient (multiple R). On the other hand, the equations also contain some significant non-linear (square) and occasionally significant interaction terms, suggesting that odor hedonics, like other taste and food hedonics, conform to a non-linear function of concentration, with a potential set of intermediate bliss points.

For these data, liking generally peaks in a middle concentration, rather than peaking at the extremes. This implies that some odorants, but not all odorants, show bliss points at intermediate levels.

Nature of the Acceptability Curve For An Odorant

We can also inquire as to the sensitivity of odor hedonics to changes in the concentration. Do all odorants, despite their different qualities, behave similarly with respect to hedonics as they change concentration? Does a 10 unit increase or decrease in concentration beyond the bliss point generate the same change in overall liking for each of the odorants.

In order to answer this question, one needs first to develop the non-linear equations as shown in Table VIII. One can now extend the concentrations outwards, by increasing or decreasing the concentration of each odorant, by a constant amount, keeping the other odorant at a fixed level near the bliss level. Table IX (A) shows this change in liking from the bliss level for one experiment, assuming various changes in concentration. The theoretical part of the analysis uses the partial derivative of liking with respect to each odorant level (Table IX (B)).

As Table IX shows, overall liking of the odorants varies as a function of the specific odorant. Each odorant shows a unique function relating concentration and liking, with this function often involving concentrations of the other odorant.

Constrained Optimization

Overall liking can be constrained in several different ways. The previous section concerned constraints in terms of concentration levels; namely, the odorant concentrations could not exceed the concentrations tested, because of possible extrapolations beyond the regions tested into regions where no data exist.

One can also constrain the odorant mixtures to maximize acceptance, while at the same time maintaining a perceptual characteristic within pre-set boundaries. Recall that overall liking or acceptability grew according to a quadratic function of odor combinations, of the form:

$$\text{Liking} = k_0 + k_1 A + k_2 A^2 + k_3 B + k_4 B^2 + k_5 AB$$

Furthermore, recall that the sensory characteristic can often be represented by a simple linear equation of components of the form:

$$\text{Sensory characteristic} = k_0 + k_1 A + k_2 B$$

In order to optimize acceptance, subject to constraints on sensory levels, we turn the problem into a straightforward optimization problem: Maximize a quadratic function (viz., liking) subject to ingredient constraints on the concentrations, and subject to linear constraints (viz., sensory characteristics).

Table X shows some typical optimization results obtained when constraining specific sensory characteristics of each mixture to lie within specified boundaries. Not all constraints work, however. The chemist, perfumer or fragrance developer must be sure that the constraints are compatible with the mixture. It does little good to constrain the sensory characteristics to lie in a region that is never reached by any feasible mixture of the odorants.

Discussion of Acceptance Optimization

These data reveal that acceptability of specific odor characteristics varies with concentration. They also reveal that the interactions of odor ingredients play a smaller role in generating acceptance of chemical mixtures than one might think. In at least the case of pairwise odor mixture, most of the variability in acceptance ratings comes from the concentration level, somewhat less from the square of concentration (allowing for an intermediate bliss point), and far less from the pairwise interaction of the odors. The contribution made by interaction might be

TABLE VIII

Non-Linear Liking Equations

Liking =	Experiment 1	Experiment 2	Experiment 3	Experiment 4
Intercept (k_o) +	11.42	21.02	14.35	16.16
k_1 (Component A)	0.163 (−78)	0.412 (.37)	0.318 (.23)	0.481 (−.21)
+ k_2 (Component A)^2	−0.000 (−.81)	−0.005 (.30)	−0.003 (.21)	−0.008 (−.26)
+ k_3 (Component B)	0.287 (−.20)	0.247 (−.01)	0.896 (.14)	0.512 (−.557)
+ k_4 (Component B)^2	−0.002 (−.59)	−0.003 (−.05)	−0.012 (.06)	−0.010 (−.62)
+ k_5 (Component A)(Component B)	−0.005 (−.23)	−0.003 (−.15)	−0.004 (.06)	−0.003 (−.53)
Multiple R	0.88	0.75	0.68	0.80
F Ratio	11.75	4.74	3.01	6.41

	Experiment 1	Experiment 2	Experiment 3	Experiment 4
A =	Isoamyl Acetate	Methyl Salicylate	Amyl Acetate	Heptyl Acetate
B =	Amyl Acetate	Ethyl Salicylate	Ethyl Salicylate	Amyl Acetate

TABLE VIII (Page 2 of 2)

Optimal Levels			
A	10.43	36.8	34.3	27.9
B	27.65	24.0	31.8	23.2
Liking at optimum	16.4	31.6	34.0	28.8
Linear R*	0.82	0.37	0.28	0.60

Numbers in parenthesis represent partial correlations

*From Table VI

TABLE IX (A)

Sensitivity of Liking to Levels of Iso-Amyl
and Amyl Acetate

Isoamyl Acetate	Amyl Acetate	Liking
10.40	23.70	15.99
12.40	23.70	15.84
14.40	23.70	15.66
6.40	25.70	16.31
8.40	25.70	16.21
10.40	25.70	16.09
12.40	25.70	15.93
14.40	25.70	15.75
8.40	27.70	16.15
6.40	27.70	16.28
10.40	27.70	16.38
12.40	27.70	15.98
14.40	27.70	15.79
6.40	29.70	16.40
8.40	29.70	16.30
10.40	29.70	16.16
12.40	29.70	15.99
6.40	23.70	16.19
14.40	29.70	15.79
6.40	31.70	16.39
8.4	31.70	16.27
10.40	31.70	16.13
12.40	31.70	15.95
14.40	31.70	15.74

TABLE IX (B)

Sensitivity of Acceptance Function to Changes in Odorant Level

Experiment 1 (Isoamyl Acetate and Amyl Acetate)

Illustration of Theory

Liking = 11.42   +0.163(Isoamyl Acetate)-0.006(Isoamyl
                            Acetate)^2
                        +0.287 (Amyl Acetate) -0.002 (Amyl
                            Acetate)^2
                        -0.005 (Isoamyl Acetate) (Amyl Acetate)

$\dfrac{\partial(\text{Liking})}{\partial(\text{Amyl Acetate})}$ =   0.287 -0.004 (Amyl Acetate) -0.005 (Iso-
                            amyl Acetate)

$\dfrac{\partial(\text{Liking})}{\partial(\text{Isoamyl Acetate})}$ =   0.163 -0.012 (Isoamyl Acetate) -0.005
                            (Amyl Acetate)

$\dfrac{\partial(\text{Liking})}{\partial(\text{Components})}$ =   Rate of change of liking per unit change
                            in odor component level

Amyl Acetate more important in changing liking than isoamyl acetate when

$$\left|\frac{\partial(\text{Liking})}{\partial(\text{Amyl Acetate})}\right| > \left|\frac{\partial(\text{Liking})}{\partial(\text{Isoamyl Acetate})}\right| \quad \text{or}$$

|0.287 -0.004 (Amyl Acetate) -0.005 (Isoamyl Acetate)| >
|0.163 -0.012 (Isoamyl Acetate) -0.005 (Amyl Acetate)|

                            or
in the simplest case:
(Amyl Acetate) - 7 (Isoamyl Acetate) > 124

(∂ = partial derivative)

TABLE X

Non-Linear Constrained Optimization of Liking

	Experiment 1		Experiment 2	
Constraint	None	Banana ≤ 20	None	Wintergreen ≤ 25
Optimal Liking	16.23	14.25	31.56	27.84
Component 1	Isoamyl Acetate 10.42	0	Methyl Salicylate 36.77	15.04
Component 2	Amyl Acetate 27.59	12.59	Ethyl Salicylate 24.00	9.51

Attributes

#	Experiment 1			Experiment 2		
1	Intensity	34	23	Intensity	39	25
2	Complexity	17	13	Complexity	28	26
3	Banana	25	20	Carnation	9	9
4	Sweet	55	13	Floral	7	8
5	Fruity	21	17	Green	10	8
6	Heavy	23	14	Spicy	18	13
7	Flowery	4	3	Minty	29	21
8	Aromatic	14	11	Sharp	21	14
9	Fragrant	11	11	Wintergreen	37	25
10	Rotten	2	1	Medicinal	15	11
11	Winey	2	1	Heavy	16	11
12	Green	2	1	Flowery	7	8
13	Herbal	6	3	Peppermint	23	16
14	Fermented	4	3	Fruity	12	11

TABLE X (Page 2 of 2)

	Experiment 3	Experiment 3	Experiment 4	Experiment 4
Constraint	None	Intensity ≈ 30	None	Heavy ≈ 15
Optimal Liking	34.0	30.3	28.8	25.9
Component 1	Amyl Acetate 34.26	8.26	Heptyl Acetate 27.9	26.3
Component 2	Ethyl Salicylate 31.76	22.76	Amyl Acetate 23.2	5.9
Attributes				
1	Intensity 41	30	Intensity 33.9	23.1
2	Complexity 35	32	Complexity 31.8	30.5
3	Fruitiness 30	29	Banana 32.8	24.6
4	Fragrant 5	3	Sweetness 21.4	17.8
5	Minty 24	20	Fruitiness 30.1	22.7
6	Banana 15	6	Heaviness 24.0	15.0
7	Sweetness 8	7	Floweriness 5.8	4.
8	Aromatic 27	21	Aromatic 21.7	14.9
9	Pear 12	6	Fragrant 16.4	12.7
10	Flowery 20	15	Pear 14.8	12.8
11	Heavy 20	14	Winey 7.7	4.4
12	Spearmint 23	20	Rotten 5.4	3.4
13	Winey 6	4	Herbal 6.8	6.2
14	Herbal 6	4	Green 1.6	1.7

higher, given complex odor stimuli, in which the subs replace single chemicals.

The data also brings up another interesting point. Odors vary widely in acceptability, as a function of type of odorant, and as a function of concentration. When researchers test acceptance/rejection of an odorant they often do so at a single concentration, without fully exploring the possibility that odors can vary in acceptance, peaking in acceptability at a middle or low range.

Another outcome of these studies is the ability to optimize accept-ability, while at the same time controlling in part the sensory "qualities profile" of the mixture. One can accomplish this by formally representing odor quality as a weighted linear combination of components, for the pairwise odor mixtures (or a weighted linear combination of more than the 2 components, for more complex mixtures). Mathematically, the con-version of odor quality of specific notes to a linear combination of concentration provides the researcher and the chemist with a means of manipulating concentrations to generate desired levels of those charac-teristics. Furthermore, within the same framework, the researcher and chemist can develop highly acceptable mixtures, with specific sensory characteristics, by constraining the sensory characteristics to lie within certain predesignated levels. The mathematical representation of odor quality in actual numerical terms makes this manipulation possible.

On The Interaction of Odor Constituents for Liking

One of the surprising outcomes of these sets of studies is the failure to find more significant interactions terms between odorants, in terms of the size of the coefficient for the interaction term, and the value of the partial correlation of the interaction term. This suggests that in such simple binary odor systems interactions may not add as much to overall liking ratings as one might expect. Rather, in the evaluation of liking the panelists assign ratings which suggest that they react to the components separately, treating each one as if it obeyed its own separate quadratic equation. The interaction term emerges, but contributes relatively little additional predictive power over and above the linear and square terms for each concentration. One would probably expect a similar under-representation of interaction terms as partial predictors of such odor qualities, such as floralness, mintiness, complexity, for binary mixtures of single chemicals. These characteristics can be fairly well modelled by means of linear equations (see Table VI). The addition of quadratic terms to each concentration will add a little more predictability. More often than not the combination of linear and square terms totally preempts the additional information to be gained by putting in yet an additional cross-term to represent the pairwise interactions of the components. Perhaps more significant pairwise interaction terms would emerge in either higher odor mixtures of 3 or more chemicals, or in truly complex mixtures, such as combinations of perfumer's subs. (i.e. mixtures which have a rose or floral quality).

It is challenging to speculate as to just precisely what occurs in the individual's mind as he or she makes the acceptance judgments. Do panelists separate out the components, and rate those components, integrating the ratings in a particular way? The panelists must be doing other things as well. Their ratings of overall acceptability often cannot be modelled as well as one can model ratings of intensity or other, more salient characteristics, even with non-linear predictors. The poorness of fit occurs with pairs of the more acceptable odorants. The goodness of fit improves when one tests combinations of an acceptable and an unacceptable odorant. Perhaps it is easier to judge a mixture of a simple pleasant odor and an unpleasant one than to judge two odorants in combination which are both pleasant, but which in context may smell too intense.

Components in Mixtures - Does The Same Chemical Behave Similarly In Different Contexts?

This set of experiments investigated several odors in different pairs. For example, Experiment I paired isoamyl acetate with amyl acetate. Experiment 3 paired amyl acetate with ethyl salicylate. One can inquire as to how amyl acetate behaves in the present of a similar smell (iso amyl acetate) vs how it behaves in the presence of a dissimilar odor (ethyl salicylate). How effective is amyl acetate in introducing its specific odor notes or changing liking when combined with iso amyl acetate as compared to combinations of amyl acetate with ethyl salicylate.

In order to answer this question let us consider the concept of relative importance of the odorant. Relative importance refers to the rate at which a sensory characteristic or a liking rating changes, per unit change in odorant concentration. In order to estimate this rate of change of characteristic per unit concentration change, one must compute the partial derivative of the sensory characteristic with respect to each odorant. (See Table IX (B)) The partial derivative is the slope, or rate of change at a point. For a linear equation, the partial derivative is a fixed number, and is given by the coefficient in the linear equation:

E.g.,: If Intensity $= k_o + k_1 A + k_2 B$

then the partial derivatives (or the rates of change of intensity with respect to A and B, respectively) are:

$$\frac{\partial (Intensity)}{\partial A} = k_1 \qquad\qquad \frac{\partial Intensity}{\partial B} = k_2$$

By comparing these partial derivatives (or in effect comparing the coefficients) for different mixtures comprising the same chemical against different background odors, one can determine the relative role which the same chemical plays in different mixture contexts.

Table XI compares the partial derivatives for common chemical components and attributes tested in the different experiments. Such

TABLE XI (A)

Relative Importance Values (Partial Derivatives)[*]
For Odorants Tested Against Different Background

Amyl Acetate	Intenstiy	Complexity	Banana	Flowery
vs Isoamyl Acetate	0.38	0.14	0.20	0.03
Vs Ethyl Salicylate	0.25	0.07	0.38	0.11
vs Heptyl Acetate	0.61	0.09	0.60	0.06
	Sweet	Winey	Herbal	Fragrant
vs Isoamyl Acetate	0.09	0.02	0.11	0.00
vs Ethyl Salicylate	0.00	0.07	0.04	0.05
vs Heptyl Acetate	0.21	0.90	0.09	----
	Heavy	Aromatic	Fruity	
vs Isoamyl Salicylate	0.33	0.10	0.20	
vs Ethyl Salicylate	0.14	0.14	-0.15	
vs Heptyl Acetate	0.73	0.66	0.55	

TABLE XI (B)

Ethyl Salicylate	Intensity	Complexity	Minty	Heavy
vs Methyl Salicylate	0.35	0.09	6.20	0.14
vs Amyl Acetate	0.47	0.03	0.25	0.30

	Flowery	Fruity
vs Methyl Salicylate	-0.01	0.03
vs Amyl Acetate	0.18	0.49

* The relative importance value = coefficient in the linear
  equation

attributes include odor intensity, liking, banana (for the acetates), minty (for the salicylates), etc. They reveal that the rate of change of a sensory characteristic or liking with respect to physical concentration of the odorant depends, to a great extent on the other odorant that is present.

This differential contribution depending on context means that when modelling qualities or liking ratings of odors in mixtures, the interaction modifies the base contribution of each odorant to the overall mixture. One cannot superimpose independent equations relating overall intensity or quality notes, for two odors each evaluated separately, and then add to that pair of independent equations an additional factor (viz., the cross term) which accounts for the mixture. This suggests that an algebra of odor mixture, with which to develop new qualities, cannot begin as an alphabet would, comprising a set of letters, which add in a simple manner, and which then entrain a third term of account for the unique pairwise interactions. Rather, the shapes and meanings of the letters or the "notes" of the component odorants change in combination, as compared to these evaluated. An odor will generate a different contribution in one odor mixture than in another. This finding bears upon the nature of the ultimate algebra of odor quality mixtures, suggesting that it will not be a simple linear one.

Discussion and Conclusions

This paper has concerned an alternative method for generating odors of specific quality profiles and acceptability levels, by mixing together simpler odorants in known concentrations. The results suggest that it may be possible to synthesize some particular mixtures, if and only if the components in that mixture produce the smell. This paper discusses mixing rules to generate predesignated sensory profiles. The profile-matching method cannot generate a new odor ab initio, unless the odor quality pre-exists in one of the mixture components.

Although not meant as a replacement for other research on odor quality, this paper suggests a possible approach to a synthesis of pre-designed odor profiles by means of mixtures of simple chemicals. The study was geared towards two component mixtures. Future studies must use a wider range of mixtures, perhaps beginning with a basic set similar to those proposed by John Amoore in his earlier work on the stereo-chemical theory of olfaction. (14) Amoore had suggested 7 primaries. Mixed together by experimental design methods (to avoid the many thousands of mixtures), these 7 basic odors might exhibit a much wider variety of qualitative nuances than can two odors ever possibly show. Statistical methods, such as the central composite design, would allow for as few as $2^7 + 2 \times 7 + 1 = 143$ mixtures. A full scale evaluation of those mixtures on attributes, coupled with profile-fitting and acceptance opti-mizations might produce much greater insight into the possibility of synthesizing predesignated odor profiles by mixing chemical components. That experiment waits for the adept chemist and psychophysicist.

References

1) Berglund, B., Ann. New York Academy of Sciences, (1974), 237, 55.

2) Berglund, B., Berglund, U. & Lindvall, T.   Acta Psychologica, (1971), 35, 255.

3) Cain, W.S., Chem. Senses and Flavor, (1975), 339.

4) Moskowitz, H.R., & Barbe, C.   Sensory Processes, (1977), 1, 212.

5) Laffort, P. & Dravnieks, A. Journale de Physiologie, (1978), 74, 19A.

6) Moskowitz, H.R., Dubose, C.N., & Reuben, M.J.   Flavor Quality: Objective Measurement, (ed. R. Scanlan).   (1977) American Chemical Society, 29.

7) Moskowitz, H.R., Chem. Senses & Flavor, (1979), 4, 163.

8) Moskowitz, H.R., Journal Food Quality, (1977), 195.

9) Moskowitz, H.R., Kapsalis, J.G., Cardello, A., Fishken, D., Maller, O, & Segars, R.   Food Technology, (1979), October, 33, 84.

10) Szczesniak, A.S. Brandt, M.A., & Friedman, H.H.   Journal of Food Science, (1963), 28, 397.

11) Beebe-Center, J.G.   The Psychology of Pleasantness and Unpleasantness, (1932) Van Nostrand Reinhold, New York.

12) Doty, R. Perception & Psychophysics, (1975), 17, 492.

13) Moskowitz, H.R., Chemical Senses and Nutrition (ed. M.R. Kare & O. Maller), (1977), Academic Press, NY, 71.

14) Amoore, J.   Olfaction and Taste III (ed. C. Pfaffmann, (1969), Rockefeller University Press, N.Y., 158.

15) Mullen, K., & Ennis, D.M. Food Technology (1979), 33, August, 74.

RECEIVED November 25, 1980.

Development of Fragrances with Functional Properties by Quantitative Measurement of Sensory and Physical Parameters

C. B. WARREN

International Flavors and Fragrances, Inc., 1515 Highway 36, Union Beach, NJ 07735

Man is blessed with the sense of smell, taste, touch, vision, and hearing. Three of these senses (touch, vision, hearing) are referred to as the physical senses and are used for detection of mechanical, thermal, photic, and acoustic energy.[1] The other two, the chemical senses, are used for the detection of volatile and soluble substances. The stimuli that excite the physical senses can be measured by both physical and psychophysical means. The volatile and soluble substances that excite the chemical senses can be defined but the stimuli caused by these substances can only be measured by psychophysical means.[2,3,4] For all practical purposes these stimuli cannot be expressed as some unit of energy, instead they have to be expressed in the dimensions of quality, intensity, duration, and like and dislike.

It is this lack of a physical method of measurement for substances that excite the chemical senses that makes the flavor and fragrance industry unique. Perfumes and flavorists are needed for the creation of its products and expert sensory panels are needed for quality control of the starting ingredients and finished formulae. Although organic and analytical chemistry are used to provide the starting ingredients and analyze finished products these disciplines cannot be used to judge quality and esthetics. There are no "iron noses" or "microprosessor tongues".

In the past five years, the quantitative measurement of quality, intensity, duration and hedonics of flavors and fragrances has become important. The measurements are used both for comparison of new products to those on the market and for substantiation of performance claims. For this last measurement the use of naive panels which reflect the opinions of the potential consumer becomes important. Examples of the types of measurements needed are: a) odor and flavor intensities of ingredients and finished products, b) substantivity of fragrances on skin and c) the effect of solvent on the odor intensity of a fragrance. Although the discipline of physical chemistry can be

0097-6156/81/0148-0057$05.25/0

used as a guide for some of these measurements it cannot replace the human nose. Physical techniques can describe absorption phenomena but not odor intensitites.

This paper addresses the quantitative measurement of odor properties using naive panels and presents methods for their selection and training. In particular, we will show how naive panelists are trained to use a magnitude estimation scale along with some typical results generated by a panel of this type.

Selection of Panelists for Magnitude Estimation Panels

Prior to admission to the training sessions all panelists are screened for their ability to perceive odor differences over a reasonable range of odors. In this way individuals with poor odor perception, those who might be partially anosmic, or those who do not care (poor motivation), are eliminated before they become members of the magnitude estimation panel. The test method used is a modification of the one developed by Gustave Carsch.[5] The test is made up of eight groups of blotters with three blotters in each group. Within each group two blotters may be the same, three may be the same, or all three may be different. There are five possible combinations for each group. The panelist is presented a ballot containing eight triangles. The apex of each triangle is identified by a letter which corresponds to a blotter containing the same letter. A typical example might be blotter group G,H,I, shown below where blotters G and I are lemon and H is lime. The panelist is instructed to put an equals sign on the leg of the triangle that connects similar smelling blotters and an "X" if the blotters are perceived to smell different. The panelist obtains one point for

each leg of the triangle that is <u>incorrectly</u> marked. If the panelist correctly discriminates between all odors he receives a zero score; if he <u>incorrectly</u> discriminates between all odors he receives a score of 24. All panelists receiving a score of nine or more are rejected.

The difficulty of the test can be adjusted by the choice of the odorants. In the extreme case, the test could be adjusted to measure the discrimination skills of a perfumer. This, however, was not our objective and the odorants chosen (see Table I), were those that a panelist untrained in perfumery might have come in contact with previously.

Magnitude Estimation Panel Training

Training starts with the magnitude estimation of the area of a series of shapes which are presented in an 18-page booklet containing a randomly sorted collection of six rectangles, six circles, and six triangles. Each page contains one figure and a 5-digit code number. The rectangles, circles, and triangles are of different sizes. The following instructions, which were adapted from a similar training exercise developed by Dr. Moskowitz,[6,7] are given to the panelists.

"Please look at the first shape in your training booklet. Do not look through the booklet, instead, pay attention only to the first shape. you are going to assign numbers that show how large the shapes you will see in the booklet seem to you. Give the first area any number you wish, write this number on the ballot sheet, along with the code number for the area. Remember, you will be using this first number to compare the size of the first shape to the size of other shapes which could be larger, smaller, or the same size as the first shape. Therefore, there are no upper limits to the size of the number you use but the number should not be so small that you cannot easily divide it into smaller portions, (smaller than 10, for instance). Now turn to the next page in the booklet. Give a number which represents the area of the shape on this page. If you give a number of 30 to the first shape and the second shape seems to the same size, give it a 30. If the second shape seems to be only one-half as large as the first shape give it a 15; if it appears to be three times as large, give it a 90. Now work through the booklet and evaluate the rest of the shapes."

Generally, this first exercise takes about 15 minutes to complete. panelists are helped if they do not understand the instructions. However, panelists who continue throughout the entire training session to not understand the instructions are rejected from the panel. Such a rejection is very rare. Table II presents a typical set of results obtained from an area estimation exercise.

Training next proceeds to estimation of hedonic tones (like and dislike). The scale for like and dislike is twice as long as that for intensity. The zero point on the scale is neither like or dislike of the stimulus, the positive side of the scale denotes like, and the negative side denotes dislike.

To obtain practice in the use of this bipolar scale the panelists are asked to magnitude estimate their like or dislike of the following words: flowers, sun, hate, worm, kiss, puppy, pollution, money, New York City, mud, perfume, murder, sex, cigar, spaghetti, rattlesnake, and love. This particular choice of words was developed by Moskowitz [6,7] to cover a dynamic range of like and dislike. Words denoting types of foods or odors also work well.

TABLE I
MATERIALS USED FOR OLFACTORY TEST

Blotter
Letter

A	South American Petitgrain Oil
B	Distilled Italian Bergamot Oil
C	Distilled Mexican Lime Oil
D	Fixateur 404, obtained from Firmenich
E	Grisalva and isomers
F	Fixateur 404, obtained from Firmenich
G	California Lemon Oil
H	Distilled Mexican Lime Oil (same as C)
I	California Lemon Oil (same as G)
K	Spanish Rosemary Oil
L	Terpineol
M	Sauge Sclaree, French
N	California Orange Oil
O	California Orange Oil
P	Grapefruit Oil
Q	Spearmint Oil
R	Spearmint Oil
S	Natural Peppermint Oil
T	Bay Oil
U	Spearmint Oil

TABLE I
MATERIALS USED FOR OLFACTORY TEST (con't)

Blotter
Letter

W	Terpeneless Lavandin
X	Distilled Mexican Lime Oil
Y	Distilled Italian Bergamot Oil
Z	Distilled Mexican Lime Oil

TABLE II

MAGNITUDE ESTIMATION OF AREAS [a]

Shape	Area (CM^2)	Estimated [b] Area	Standard [c] Error
Circle	7.1	7.8	1.11
	19.7	10.0	1.07
	43.0	31.1	1.05
	91.6	52.2	1.05
	145.3	69.5	1.05
	216.4	106.9	1.04
Triangle	2.0	4.7	1.04
	7.4	10.7	1.11
	24.8	22.5	1.07
	64.9	49.7	1.04
	104.0	65.4	1.04
	322.0	92.4	1.03
Square	10.1	13.2	1.10
	17.9	18.8	1.07
	72.4	47.6	1.03
	123.0	68.5	1.03
	123.0	69.5	1.03
	203.0	97.4	1.02

a.   The sequence of the areas presented to panelists was random. The results were sorted by shape and size for this table. Twenty two panelists were used for this exercise.

b.   Estimated areas were normalized by the averaging method. The values presented in this table are geometric means.

c.   The standard error is for the geometric mean and equals 1 + percent errors.

The following set of instructions adapted from those of Moskowitz are used to introduce the hedonics training session.

"As another exercise we would like you to express your liking or disliking of different words. Using the bipolar scale discussed previously, show how you feel about each word. If you like a word write an L next to it. If you dislike the word, write a D next to it. Then indicate how much you either like or dislike the word by also writing in a number. A large "L" number means you like it a lot, while a large "D" number means you dislike it a lot. On the other hand, a small "L" number means you like it a little, while a small "D" number means you dislike it a little. If you feel indifferent or neutral about a word, give it a zero (0). As an example, suppose you gave the first word an "L 10" to show how you felt about it but you like the second word twice as much. The second word should receive a score of "L 20". If you dislike the third word, you should give it a "D" and a number to show how much you dislike it. If you dislike it a lot you might give it a "D 100". Remember that the particular scale you use is your own. There are no limits to the size of the scale and no one's scale is more right than any one elses."

Table III presents a typical training panel's hedonic scores for the 17 words discussed previously. Although the panel was asked to use "L" and "D" to denote like and dislike, the scale is actually positive numbers for like and negative numbers for dislike. Our experience has shown that the panelist can use L and D with much less difficulty than plus and minus.

The final task of the training session is the tasting or smelling of samples. The choice of the samples generally depends on the first evaluation task to be carried out by the newly trained panel. Thus, the panel that evaluated hydrolyzed vegetable protein tasted a concentration series of glucose solutions for their training session. Whereas, the panel that self-evaluated underarm odor smelled a concentration series of synthetic body odor in their training session. Table IV presents glucose flavor intensities and hedonics obtained during a training session by the same 22-member panel that provided the shape and word evaluations presented previously.

So far over 100 members of the R&D staff at International Flavors & Fragrances have been trained to magnitude estimate odors and flavors. The complete training session takes about one and one-half hours and has been used to train secretaries, engineers, managers, chemists, maintenance workers and clerks. The data presented in this paper were obtained by these people.

TABLE III

MAGNITUDE ESTIMATION OF THE LIKE AND DISLIKE
FOR A SERIES OF WORDS.[a]

WORD	HEDONIC SCORE	STANDARD ERROR
Sex	138	15.8
Love	126	10.1
Kiss	84	8.9
Money	82	7.0
Sun	73	6.8
Flowers	58	7.0
Puppy	47	6.1
Spaghetti	41	6.7
Perfume	37	5.1
New York City	0	13.7
Worm	-5	6.4
Mud	-25	4.0
Cigar	-31	12.8
Rattlesnake	-40	11.4
Pollution	-72	6.9
Hate	-81	10.9
Murder	-140	17.3

a.   The words presented in Table III have been sorted on a
     like-dislike scale. The sequence of the words presented to
     the panel was in random order. The results presented here
     were obtained with a 22-member panel.

TABLE IV
FLAVOR INTENSITY AND HEDONICS OF GLUCOSE SOLUTIONS.

CONCENTRATION IN WATER (%)	INTENSITY	STANDARD ERROR	HEDONICS	STANDARD ERROR
2	5.8	1.2	1.6	2.3
4	14.5	2.0	5.1	3.2
8	32.7	3.0	8.4	6.6
16	58.5	4.7	-1.0	6.5
32	112.3	4.5	-30.7	10.2

Analysis of Magnitude Estimation Data.

As you remember, panelists were told only to choose numbers so that the ratios of the numbers reflected the ratios of their perceptions. The choice of the particular range of numbers was left up to the panelist. In order to eliminate the variance due to scale differences magnitude estimation data need to be normalized.[8]

Normalization is a technique in which each panelist's evaluation is multiplied or divided by a factor which transforms it to a common scale. This paper presents an averaging and an internal standard method of calibration which was used for the data presented herein. Also commonly used is an external calibration method which is described in ref. [6].

Averaging Method

This method can be used for normalization of hedonic as well as intensity data. The first step is the determination of the magnitude of the scale used by each panelist by summing the absolute values of all of his or her evaluations for a particular panel session.

$$\text{Panelist's Scale Magnitude} = \sum_j \left| X_{ij} \right|$$

X_{ij} = the numerical evaluation

for the ith panelist.

over the j evaluations.

The second step is the calculations of the scaling factor for the particular panel session by summing the absolute values of all evaluations of all panelists and dividing by the number of panelists.

$$\text{Panel Scaling Factor} = \left(\sum_{i}^{n} \sum_{j} \left| X_{ij} \right| \right) \Big/ n$$

i = panelist index
j = evaluation index
n = number of panelists

The correction factor for each panelists is calculated with the equations presented below.

$$\text{Correction Factor} = \frac{\text{Panelist's Scale Magnitude}}{\text{Panel Scaling Factor}}$$

Internal Standard Method

This type of normalization procedure works well for measurement of odor intensities. We have chosen the use of 270 parts-per-million of n-butanol in water as the internal standard for odor intensity evaluations.[9] The current procedure is to place three butanol-in-water standards into a typical sample set made up of 20 samples. Standards and samples contain 5-digit random number codes, the sequence in which each panelist smells the samples and standards is completely random. The correction factor for a particular panelist is the constant that will adjust the average of the perceived intensities for the butanol samples to 30. The sample intensities obtained by this panelist are then normalized by multiplication by the correction factor:

Example:

Panelist's Scale Magnitude = $\overline{X}_i$ (butanol)

where $\overline{X}_i$ (butanol) is the average intensity for the three butanol samples for the ith panelist

Correction Factor = $f_i = 30/\overline{X}_i$ (butanol)

$$\hat{X}_{ij} = X_{ij} \cdot f_i$$

where $\hat{X}_{ij}$ = the normalized intensity of sample j for panelist i.

Use of Magnitude Estimation Results.

Dose Response Curves.

Some 25 years ago S. S. Stevens[10] found that sensory data generated on a magnitude estimation scale could be fitted to a power function such as the one presented below.

$$\text{Magnitude Estimation} = \underline{a}\ (\text{stimulus})^b$$

Two examples of this fit are shown below for the area and the glucose intensity data presented previously in this paper.

$$\text{Estimated Area} = 2.77\ (\text{Area})^{0.66}$$

$$r^2 = 0.99$$

$$\text{Estimated Glucose sweetness} = 3.7\ (\text{concentration})^{1.05}$$

$$r^2 = 0.99$$

Dose-response curves have been used by the fragrance indus-try to describe odor intensities of aroma chemicals and perfumes in the concentration range of their use.   The curves have been valuable for the comparisons of the relative odor intensities of aroma chemicals in the same odor class and for measurement of the effect of solvent on odor intensity.   Examples of some compari-sons are presented in Table V, chemical structures are presented in Table VI. Galaxolide and indisan, for example, have slightly flatter intensity curves than Musk Ambrette or Sandiff.   These data suggest that galaxolide or indisan will have higher odor intensities at lower concentrations than will the corresponding odorants.   Knowledge of the complete equation allows one to calculate odor intensities at any concentration within the concentration range of the measurements.   Table VII shows the dose-response exponents for three fragrances and two aroma chemicals in diethyl phthalate and in a less polar solvent. These data suggest that the less polar solvent tends to flatten the intensity curve, that is, the solvent swallows up the fragrance.   Another interesting aspect of the data is the decrease of r^2 for galaxolide and indisan in the new solvent. This indicates that only about 60 to 70 percent of the variation of the perceived odor intensity is due to its variation in concentration suggesting that the solvent is donating part of the odor.

Correlation of Physcial With Phychophysical Measurements.

In general, a psychophysical measurement is more expensive and more tedious to obtain than a physical measurement. Compare, for example, the time and expense required to measure quantita-tively an odor recognition threshold for a particular molecule vs

TABLE V
DOSE-RESPONSE CURVES EXPONENTS FOR VARIOUS AROMA
CHEMICALS.[a,b]

Compound	Exponent	r^2
Musk Odorants:		
Galaxolide	.27	.95
Musk Ambrette	.34	.93
Sandalwood Odorants:		
Indisan	.30	.94
Sandiff	.44	.99
Some Other Odorants:		
Methyl Ionone, Gamma A	.17	.96
Lyral	.24	.93
Cinnamalva	.34	.99
Isocyclemone E	.47	.96

a. All materials were dissolved in DEP and measured in a
concentration range of 0.2 to 20%.

b. Structures are presented in Table VI.

TABLE VI
STRUCTURES OF MOLECULES USED FOR DOSE-RESPONSE CURVES.

COMPOUND	STRUCTURE
Galaxolide	
Musk Ambrette	
Methyl Ionone, Gamma A	
Lyral	
Cinnamalva	
Isocyclemone E	

TABLE VII
USE OF DOSE-RESPONSE CURVES TO COMPARE SOLVENTS.

MATERIAL[a]	DEP [b] Exponent	r^2	LESS POLAR SOLVENT [c] Exponent	r^2
FRAGRANCE A (Citrus, Coumarinic Woody and Sweet)	.26	.92	.12	.98
FRAGRANCE B (Heavy, Woody, Floral with strong patchouli note)	.33	.90	.16	.91
FRAGRANCE C (Spicy, floral)	.43	.93	.14	.96
GALAXOLIDE	.27	.95	.11	.63
INDISAN[d]	.30	.94	.11	.70

a. Concentration range for dose-response curves was 0.20 to 20%.

b. DEP is diethyl phthalate.

c. This solvent is less polar than DEP.

d. Indisan is the product name for a complex mixture which has a sandalwood odor.

the time and effort to obtain the infrared absorption spectrum
for the molecule. The former measurement requires the time of 10
or more panelists, the preparation of a number of solutions of
the molecule at different concentrations and the work-up of the
data. The latter measurement requires one solution, one person,
and about 10 minutes for the scanning of the infrared spectrum.
This difference in time and money has led us to develop physical
methods of measurement that compliment the psychophysical
methods.

One such area was the measurement of the detergent powder
fragrance retained by cloth at the end of a laundry wash cycle.
There are two ways to perform such a requirement. One can use
sensory panels to measure the retention of either a finished
fragrance or individual aroma chemicals on cloth, or one can
develop a physical method for measurement of the concentration of
aroma material on the fabric surface. We have developed methods
for such a measurement by use of partition coefficients and
Tables VIII and IX present some representative data. The
physical meaning of the partition coefficients presented in these
tables is the following:

$$K = \frac{\text{Cloth Concentration of Aroma Chemical}}{\text{Wash bath Concentration of Aroma Chemical}}$$

A partition coefficient of zero indicates that none of the
aroma chemical is on the cloth. A partition coefficient of one
indicates equal distribution between cloth and wash-bath. The
larger the partition coefficient the higher the affinity of the
material for the cloth. Both Tables present partition coefficients vs odor intensitities of the aroma chemical or
fragrance: 1) on the detergent powder, 2) above the wash water
during the wash cycle, 3) on the cloth after two rinse cycles,
and 4) on the cloth after two rinse cycles and hot air drying.
Analysis of the partition coefficients versus the perceived odor
intensities presented in Table VIII suggest the following:

1) Acetophenone has a high odor intensity and a low
   partition coefficient, thus it will have a high odor
   intensity on the detergent powder but a relatively low
   odor intensity on cloth since it prefers to stay with
   the aqueous phase.

2) Musk Ambrette has a low odor intensity and a high
   partition coefficient, thus it will have a relatively
   low odor intensity on the detergent powder and a high
   odor intensity on the wet and dry cloth.

TABLE VIII
COMPARISON OF PERCEIVED ODOR INTENSITIES
WITH PARTITION COEFFICIENTS.

MOLECULE	K[a]	ODOR INTENSITIES [b]			
		POWDER	WATER	WET CLOTH	DRY CLOTH
ACETOPHENONE	.5	126	17	9	5
CINNAMALVA	3.3	118	8	7	5
METHYL IONONE GAMMA A	8.7	68	18	38	13
ISOCYCLEMONE E	11	31	12	24	7
MUSK AMBRETTE	17	37	12	29	12

a. K is the partition coefficient which equals concentration of aroma chemical on cloth divided by concentration of aroma chemical in the wash bath.

b. Powder = represents the odor intensity of the molecules on the detergent powder.

Water is the odor intensity of the molecule above the aqueous wash bath.

Wet Cloth is the odor intensity on the cloth after two rinses.

Dry Cloth is the odor intensity on cloth after two rinses and drying.

TABLE IX
USE OF PARTITION COEFFICIENTS TO CREATE
SUBSTANTIVE FRAGRANCES.

FRAGRANCE	K	INTENSITIES [b]			
		POWDER	WATER	WET CLOTH	DRY CLOTH
PARTITA 1	1 - 4	107	23	10	5
PARTITA 2	5 - 11	70	19	32	13
PARTITA 3	5 - 11	60	23	36	16
PARTITA 4	- 12	92	27	35	20

a. K is the partition coefficient and is defined in Table VIII.

b. Powder, Water, Wet Cloth, and Dry Cloth are defined in Table VIII.

Table IX presents the practical applications of the partition coefficient concept; that is, fragrances created from aroma chemicals with larger partition coefficients show higher odor intensities on both the wet and dry cloth. This is easily seen by comparing the partition coefficients and odor intensities for the fragrance called partita 1 to those for partita 4.

The sensory observations obtained for this detergent work were normalized by the internal standard method against 270 ppm butanol in water. Thus, odor intensities of 30 are moderate and intensities of 60 are strong.

Odor Masking.

One of the largest uses of fragrance is to mask malodors of personal and household products. Also, the general area of odor masking and blending is very important commercially[11] and academically.[7] In spite of the large amount of work in this area the literature did not contain a simple quantitative method for measurement of the masking ability of a fragrance. One solution to this problem was to magnitude estimate the odor intensity and hedonics of the fragrance plus base at several concentrations of the fragrance. Some typical examples are presented in Table X. The indication that the fragrance is either masking or improving the quality of the odor is shown by a significant increase in hedonics; accompanied by a small increase in odor intensity. (The best possible situation would be a significant increase in hedonics accompanied by a decrease in odor intensity). Table X shows that fragrance 1 is better than fragrance 2 for the latex paint while fragrance 3 provides no significant masking of the oil-base paint odor.

The Future of Magnitude Estimation (ME).

At present, the use of ME ratio scaling is both in a state of expansion and critical evaluation. The technique has been found to serve well for attitude evaluations (such as the impact of an advertisement). ME in combination with a response surface experimental design[12] has been used for optimization of food products. Ratio scaling is still experimental in that a best normalization method has not been found, nor has the method received a critical comparison to the more popular category scaling method. Both of these questions are now being addressed by the American Society for Testing and Materials (ASTM) Committee E-18 - Sensory Evaluations of Products and Materials.

Proponents of ME claim the method to be easy to teach to naive panelists, very sensitive for measurement of intensities in the supra-threshold region and very efficient for measurement of product preference relative to some bench mark. The future of the method will depend on how it stands up to a critical comparison with category scaling methods.

TABLE X
MEASUREMENT OF ODOR MASKING.

MATERIAL	FRAGRANCE CONCENTRATION (%)	HEDONICS	INTENSITY
Latex Paint & Fragrance #1	0	3	9
	0.0032	6	1
	0.01	12	13
	0.032	23	17
Latex Paint & Fragrance #2	0	3	9
	0.0032	3	14
	0.01	7	29
	0.032	5	40
Oil-base Paint & Fragrance #3	0	-17	45
	0.030	-18	60
	0.10	-16	55
	0.32	-14	75

LITERATURE CITED:

1) American Society for Testing and Materials Special Technical Publication No. 433. "Basic Principles of Sensory Evaluation", Published by ASTM; 1916 Race Street, Philadelphia, Pennsylvania 19103 (1968).

2) L. B. Sjostrom, S. E. Carincross and J. F. Cauley, "Methodology of the Flavor Panel", Food Technology 11, 20 (1957).

3) American Society for Testing and Materials Special Publication No. 434. "Manual on Sensory Testing Methods", Published by the ASTM; 1916 Race Street, Philadelphia, Pa. 19103 (1968).

4) D. R. Peryan and F. J. Pilgrim, "Hedonic Scales of Measuring Food Preferences", Food Technology 11, 9 (1957).

5) G. Carsch, "An Olfactory Aptitude Test for the Selection of a Perfume Panel". Published in Proceedings of the Second Open Symposium of the American Society of Perfumers. Available from the American Society of Perfumers, Inc., 630 Fifth Avenue, New York, N.Y. 10020 (1956).

6) H. R. Moskowitz, "Magnitude Estimation: Notes on What, How, When, and Why to Use It." Available through MPI Sensory Testing, 770 Lexington Avenue, New York, N.Y. 10021 (1977).

7) H. R. Moskowitz, "Odors in the Environment. Hedonics Perfumery and Odor Abatement", in handbook of Perception, Vol. 10; E. C. Carterette and M. P. Friedman, eds., Academic Press, New York, N.Y. (1978).

8) R. B. McCall, "Fundamental Statistics for Psychology", 2nd ed., Harcourt Brace Jovanovich, Inc.; New York, N.Y. (1975).

9) American Society for Testing and Materials Standard No. E544-75. "Standard Recommended Practices for Referencing Suprathreshold of Odor Intensity", published by ASTM, 1916 Race Street, Philadelphia, Pa. 19103 (1975).

10) S. S. Stevens, "On the Brightness of Lights and the Loudness of Sounds". Science, 118, 576 (1953).

11) A good leading reference in this area is: "Industrial Odor Technology Assessment", P. N. Cheremisinoff, R. A. Young, eds., Ann Arbor Science Publishers, Inc., P. O. Box 1425, Ann Arbor, Michigan 48106.

12) H. R. Moskowitz, J. W. Chandler, D. W. Stanley, "Eclipse™ Developing products from Concepts in Consumer Ratings." <u>Food Product Development</u>, March (1977).

RECEIVED December 2, 1980.

Sensory Structure of Odor Mixtures

ANDREW DRAVNIEKS[1], FREDERICK C. BOCK, and FRANK H. JARKE

IIT Research Institute, Chicago, IL 60616

An ultimate objective in the resolution of relations between odors and odorant structures is to predict odor from chemical identities and concentrations of odorants in air containing mixtures of odorants.

Even the first step, prediction of odor quality from the molecular structures of single odorants, is as yet uncertain. Some odor/structure relations have emerged gradually from studies by many researchers, but a comprehensive coherent theory of the structural basis of odors does not yet exist. Wherever relationships appear to exist, they are far from applicable to mixtures of odorants.

However, the relationship between the odors of single odorants and their mixtures can be investigated without regard to the molecular structures of these odorants. The sensory structures of the odors of single component odorants can be characterized, e.g., by multidimensional scaling. The sensory structure of an odorant mixture can also be characterized by some means, and then rules can be explored which tie the odor of the mixture to the odor of components.

As an example, if odorants with similar level of spicy note are mixed, what will be the spicy level of the mixture?

This approach was studied using vapor mixtures of 28 odorants, with up to 4 odorants per mixture.

Experimental

Odorants. Twenty-eight odorants covering a large variety of odor character notes and a broad hedonic tone range (from isovaleric acid to vanillin) were used:

---

[1] Current address: Institute of Olfactory Sciences, Park Forest, IL 60466.

0097-6156/81/0148-0079$05.00/0

	Initially Considered Essentially Pleasant		Initially Considered Essentially Unpleasant
A	Amylbutrate	3	Ammonia
B	Benzaldehyde	R	1-Butanol
C	Eucalyptal	Y	Butyric
J	Cinnamaldehyde	F	2,4-trans-trans
W	Citral		Decadienal
K	Coumarin	D	Diacetyl
E	Eugenol	Z	2-Ethyl-3,6-dimethyl-
G	Guiacol		pyrazine
N	Limonene	Q	Ethylsulfide
L	Linalool	H	1-Hexanal
M	Methanol	2	Hydrogen sulfide
X	Musk, pentadecanolide	I	Isovaleric acid
U	γ-Undecalactone	0	2-Octanone
V	Vanillin	P	Phenol
		S	Propylmercaptan
		T	Trimethylamine

Statistical Design. To keep the complexity of mixtures manageable, only binary, tertiary, and quaternary mixtures were considered, assuming that the quaternary complexity should begin to
reflect rules operative in multicomponent mixtures.

A fractional factorial statistical design known as balanced
imcomplete blocks with separable replicates were utilized. In
each session, four odorants at a time are evaluated:

    4 odorants separately

    6 (all possible) binary mixtures

    4 (all possible) ternary mixtures

    1 (in duplicate) quaternary mixtures

In a block of 7 sessions, each odorant is used in one of the
sessions. Nine blocks (63 sessions) would include each odorant 9
times, each possible pair once, and include 1/13 of all possible
ternary and 1/325 of all possible quaternary mixtures. Because of
practical limitations, only 4 blocks could be completed, covering
168 binary, 112 ternary and 28 quaternary mixtures, with a duplication of each quaternary mixture in the same session. The design
permits statistical analysis of separate blocks.

Apparatus. Figure 1 represents the mixture olfactometer used
in the study. The apparatus consists of 16 stimuli mixing manifolds. Air at 0.5 L/min to each manifold is supplied through
stainless steel capillary tubings from the air distributor manifold; the 17th capillary branch serves to monitor the air supply
rate.

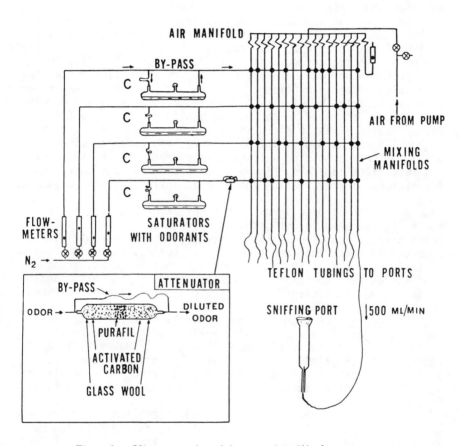

Figure 1. Olfactometer for mixing up to four (4) odorant vapors

The liquid odorants are supplied by vaporizing these in the saturators into dry nitrogen stream. In each saturator assembly, part of the nitrogen by-passes the saturator; another part, controlled by calibrated stainless steel capillary C, passes over the odorant, becomes saturated with its vapor, and mixes with the by-pass nitrogen before flowing to the mixing manifolds.

For some odorants, the needed dilution cannot be easily reached by dilution in the by-pass system alone. For these, an attenuator shown in the insert at lower right was used.

Three of the odorants (ammonia, hydrogen sulfide, and trimethylamine) were gases. Their dilutions were prepared in thick wall collapsible 18-L containers, injecting by syringe the needed amount of the odorant gas and filling with air. The dilutions were prepared one day before their use, to allow time for a stabilization after adsorption on the walls. The diluted vapors were then supplied to the mixture olfactometer by a peristaltic pump.

Stimuli prepared in the mixing manifolds were supplied by Teflon tubing lines to glass sniffing ports which had 25 mm x 35 mm elliptic opening. The ports were hung randomly along the walls in three adjoining well-ventilated rooms. The first left and last right manifold, Figure 1, supplied the same four-component mixture, for evaluating the reproducibility of the judgements.

Selection of Dilutions. A Butanol-vapor odor intensity scale (1) was used to estimate the odor intensity of stimuli consisting of single odorants. The dilutions were empirically adjusted to match the odor intensity of butanol vapor in the 50 to 100 ppm (v/v) concentration range, but in actual tests some values fell somewhat outside this range. The corresponding intensity was sufficient for clearly discerning the odor character.

Procedure. In each session, 9 panelists were used, drawn from a pool of 15, since in this several month long experiment a constant panel composition was impractical. However, in each session, all 16 stimuli were evaluated by the same panelists, so that differences between panelists, as far as odor of mixtures vs. odor of components are concerned, were not a directly complicating factor.

The mixture olfactometer was set in operation 1-2 hours before the panel session. Panelists circulated among the sniffing ports and characterized the odor quality of the stimuli using a 136-descriptor multidimensional scale, described elsewhere (2); it is an extended Harper's scale. (3)

After the session, the olfactometer was flushed with air for 1-2 days, to remove adsorbed traces of odorants.

Reproducibility. There were 28 quaternary mixtures tested in duplicate. These duplicates were evaluated the same session. Twelve descriptors were selected for testing the reproducibilities;

sour, oily, putrid, rancid, stale, burnt, sharp, bitter, herbal, ethereal, sweet, and fragrant.

For each descriptor a Chi-square test was designed. Descriptor scores were grouped in 3 classes; 0, 1 + 2, and 3 + 4 + 5, and distribution of panel responses by classes for the first and the second presentation of the quaternary stimulus, over the entire set of 28 quaternary, was compared. Chi-square values were obtained that showed a high similarity of distributions, considerably in excess of 10 percent probability. Thus, duplicate quaternary mixtures produced well-correlated descriptor responses, at least for the 12 descriptors selected for this test.

Differences in the odor of pairs of stimuli were also estimated by using a method based on coefficients of association between individualized (by panelists) multidimensional profiles.(2) The negative natural logarithm of this coefficient was previously found to correlate to the sensory distance between two profiles.

In the present study, the 28 quaternary mixtures were evaluated in duplicate. Most sensory differences within duplicates had -ℓn (coeff. assoc.) below 1. Most differences between single odorants (168 pairs) were above this value. When the value of 1 was experimentally taken as the dividing datum between "same" and "different" odor, the separation of these groups, by a Chi-squared test, was highly statistically significant. Thus, odor differences between duplicated quaternaries were at most only as large as for odorants with odors that appear to a not-highly-trained perfumer somewhat alike (citral/limonene; butyric/isovaleric acids; hydrogen sulfide/ethyl sulfide; trimethylamine/butyric acid?).

Results and Discussion

The objective of the data analysis was to discover how odors of mixtures related to the odors of components. This may be possible by comparisons of entire multidimensional profiles of mixtures and components, but such an approach requires assumptions on the appropriateness of selecting some specific profile comparison method. The complexity of rules that seem to govern the odor quality of even simple mixtures has been pointed out by Moskowitz, et al. (4)

Instead, a method was selected in which scores for specific descriptors for the components and mixtures were compared. A frequency-of-use histogram for the descriptors indicated that for the 28 odorants selected, and their mixtures, 30 descriptors were most commonly used. These descriptors are listed in Table 1. Further data analysis was confined to these 30.

Classification of Mixing Effects. For each of the descriptors, the score for a mixture can be compared to the scores of the components (concentrations of components are essentially the same for single components and these components in the mixture). Three benchmarks can be derived from the component scores: the lowest

TABLE I. MODELS FOR RELATION OF MIXING EFFECT CODE TO MEAN OF COMPONENT SCORES BY DESCRIPTORS

| DESCRIPTOR | SIMPLE REGRESSION MODEL | | | MODEL WITH FOUR SUBSIDIARY VARIABLES ADDED | |
	INTERCEPT	SLOPE	DEGREE OF DETERMINATION	DEGREE OF DETERMINATION	NAMES OF SUBSIDIARY VARIABLES
Sharp	3.6	0.67	0.48	0.58	Min. hed./Sweet/Bitter/Etherish
Heavy	3.6	0.77	0.55	0.67	Min. hed./Light/Ethereal/Powdery
Sickening	3.6	0.66	0.58	0.72	Min. hed./Medicinal/Herbal/Light
Oily	3.7	0.63	0.41	0.49	Stale/Sweet/Max. hed./Min. hed.
Sweet	3.4	0.75	0.60	0.71	Cool/Max. hed./Oily/Sour
Sour	3.6	0.67	0.52	0.64	Min. hed./Sweet/Sickening/Herbal
Stale	3.6	0.67	0.48	0.57	Min. hed./Oily/Putrid/Fruity (other)
Bitter	3.7	0.84	0.51	0.61	Min. hed./Cool/Warm/Ethereal
Warm	3.5	0.81	0.46	0.55	Clove/Woody/Max. hed./Sour
Aromatic	3.6.	0.72	0.60	0.68	Max. hed./Stale/Fragrant/Burnt
Rancid	3.7	0.75	0.56	0.71	Min. hed./Sweet/Oily/Lemon
Cool	3.5	0.77	0.43	0.83	Burnt/Max. hed./Ethereal/Disinfectant
Burnt	3.7	0.76	0.52	0.60	Fruity (other)/Herbal/Oily/Ethereal
Spicy	3.8	0.60	0.47	0.56	Clove/Max. hed./Heavy/Fruity (other)
Putrid	3.7	0.64	0.57	0.68	Min. hed./Rancid/Sweet/Burnt
Woody	3.8	0.66	0.31	0.43	Disinfectant/Aromatic/Putrid/Bitter
Fragrant	3.3	0.76	0.65	0.73	Max. hed./Minty/Fruity (other)/Lemon
Etherish	3.8	0.74	0.51	0.55	Fruity (citrus)/Lemon/Sweaty/Putrid
Minty	3.9	0.61	0.41	0.53	Burnt/Max. hed./Spicy/Aromatic
Soapy	3.7	0.81	0.41	0.50	Burnt/Lemon/Putrid/Clove
Fruity (other)	3.7	0.96	0.50	0.56	Clove/Oily/Stale/Sweet
Clove	3.9	0.60	0.44	0.51	Sour/Max. hed./Minty/Fruity (other)
Disinfectant	3.8	0.82	0.39	0.48	Woody/Putrid/Etherish/Soapy
Sweaty	3.7	0.78	0.47	0.66	Min. hed./Sweet/Rancid/Bitter
Herbal	3.7	0.60	0.37	0.44	Max. hed./Sweet/Sour/Medicinal
Medicinal	3.7	0.63	0.51	0.63	Putrid/Burnt/Stale/Oily
Lemon	3.8	0.60	0.53	0.61	Burnt/Stale/Herbal/Putrid
Powdery	(3.3)	(1.03)	0.55	0.64	Spicy/Fruity (citrus)/Minty/Max. hed.
Fruity (citrus)	3.8	0.66	0.57	0.61	Burnt/Powdery/Soapy/Herbal
Light	(3.0)	(0.80)	0.37	0.53	Fragrant/Stale/Max. hed./Herbal
Mean*	3.68±	0.71			
Standard Deviation	0.14	0.09			

* Excluding "light" & "powder"

panel mean score, the highest panel mean score, and the arithmetic mean of the component scores.

In principle, in some cases, the score for the mixture may be higher than the highest component score. Some form of the odor note additivity or a promotion by other different notes originating in the other components of the mixture may then be suspected. In the other extreme, the score for the mixture may be lower than the lowest component score, and a suppression or a dilution of the descriptor-characterized note by other notes offered by the other components may have occurred. For the inbetween cases, less pronounced effects of a similar type may have occurred.

The phenomenology of the mixing effects was inspected using the histograms in Figure 2. The mixing effect codes were as follows:

> 2 = score for the mixture is equal or lower than for the component with the lowest score.

> 3,4 = score for the mixture is higher than the lowest component score, but lower than or equal to the mean score of the components.

> 5,6 = score for the mixture is higher than the mean component score, but lower than or equal to the highest component score.

> 7 = score for the mixture is higher than the highest component score.

For orientation:

> (a). all cases with code 6 or below (left of arrows) demonstrates either suppression or at most non-impairment (if scores of mixture equal to highest component score) of the odor note upon mixing.

> (b). all blackened bars indicate cases where mixing reduced the scores to values below the mean score of the components (or, in rare cases, kept it at the mean score level).

An inspection of Figure 2 leads to the following conclusions:

> (1). There are only a very few cases where mixing might have enhanced an odor note

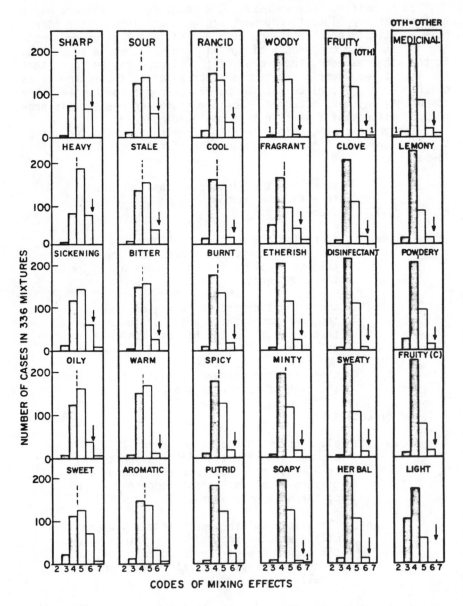

Figure 2.   *Influence of mixing on scores for 30 most frequently used descriptors*

of that component which was the highest
in this odor note--see low frequency of
occurrence of Code 7 effect.

(2). There is only one case (clove) where
mixing brought the note below the level
occurring in the component with the
lowest level of this note.

(3). Overall, the scores for the mixtures
gravitate around the mean score of the
components.

(4). Two types of contrasting behavior and a
third one of the intermediate type seem
to appear. In one, (10 left histograms),
the odor notes are retained on mixing,
remaining higher than the mean score
level. In the other, most of 10 histograms
on the right, the odor notes appear to
be more susceptible to a degradation by
mixing. "Light" is the extreme example
of the latter behavior, but it is easily
understood since this is more of an odor
intensity than quality descriptor, and
the odor will be stronger and "heavier"
as other components are added.

(5). Superficially, less specific descriptors
appear to belong to the first group, and
descriptors for more specifically recog-
nizable odor notes belong to the second
group.

Thus, the principal effect upon mixing appears to be a reduction of scores for various odor notes from the level of the score
for the most highly scored component. Odor notes also appear to
differ in their resistance to such degradation. Apparently,
introduction of other odor notes on mixing usually weakens the
level of the odor notes of the components in an analogy to the
role of an auditory noise in sound recognition.

Simple Mathematical Model for Odor Mixtures. Since the data
in Figure 2 indicated the mean of scores of the components may
serve as a crude benchmark for deriving the score for the mixture,
a mathematical model was devised for a more refined relation
between the component and mixture odor notes. The model is based
on a linear regression:

[CODE VALUE] = [INTERCEPT] + [SLOPE] [MEAN COMPONENT SCORE]

Such equations were sought for all 30 of the most frequently occurring descriptors: Table I lists the values of intercepts, slopes, and coefficients of determination (a measure of the goodness of fit to the obtained equation).

Typically, about 50 percent of variance was accommodated by such a simple equation. For most descriptors, the intercept and slope coefficients do not vary much with the descriptor. Coefficients for "light" and "powdery" are different from those for other descriptors. If these are disregarded, and the mean values of the coefficient taken, the following equation results:

$$[CODE\ VALUE] = 3.7 + 0.71\ [MEAN\ COMPONENT\ SCORE]$$

Improvements to the Model. Since other odor notes undoubtedly influence the scores of some selected odor notes, additional variables were added to the simple regression model above on a multiple stepwise regression analysis was conducted. For each odor note, the other candidate variables were all other 29 descriptor scores, and the hedonic tone of the hedonically lowest (least pleasant or most unpleasant) and highest components. (5) Only 4 subsidiary variables were allowed to enter the equation.

The last two columns illustrate the performance of the improved model. The degrees of determinations are significantly higher, but in the best case were at 0.7 level (fragrant, sickening, sweet, rancid). The four subsidiary variables for each descriptor are listed in the last column.

Procedure for Estimating Score for Mixtures; Example. Three odorants, A, B, and C, are mixed in the vapor phase. Their scores (mean panel values) for some selected descriptor are 1.8, 2.6, 3.2. The 1.8 is the lowest, corresponding to Code 2. The mean 2.53, corresponding to Code 4. The highest is 3.2, corresponding to Code 6. These three points are plotted in Figure 3.

The estimated code value for the mixture is, from the regression equation above (generalized form):

$$[CODE\ VALUE] = 3.67 + 0.71 \times (2.53)\ =\ 5.47$$

Reading back from the code value 5.47 via plot of Figure 3, the best estimated score for the mixture, for this descriptor, is 3. Note that the standard deviation for the simple regression equations of Table I typically is 0.5 on the code values.

Summary and Conclusions

Odor quality (character) of 336 mixtures of 28 odorants, up to quaternary in complexity, was evaluated using multidimensional scaling and compared with that of the component odorants.

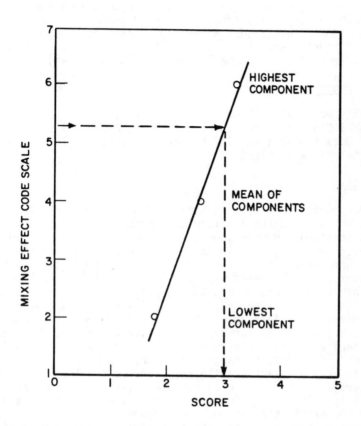

Figure 3. Example for estimating the score for mixture from scores of its components

For 30 most frequently encountered (in this work) odor notes, the odor note scores for mixtures were most frequently lower than for the component with the highest score, and most typically were close to the mean of the component scores. An enhancement of an odor note by mixing was infrequent. A suppression of an odor note to, or below the lowest component score was also infrequent. Those notes which were more specific seem to be more susceptible to degradation mixing.

Linear regression equations anchored to the mean of the component scores typically accounted for 50 percent of variance. Introduction of other odor notes and hedonic data to expand these by 4 additional variables increased the accounted for variance by about 10 percent; occasionally more or somewhat less.

Acknowledgement

This material is based on experimental work supported by S.C. Johnson & Son, Inc., Philip Morris, Inc., Procter & Gamble Co., Quaker Oats Company, and IIT Research Institute. Their support is gratefully acknowledged.

Abstract

Presumably, the relations of odor quality (character) of single odorants to their molecular properties will be eventually well-understood. However, most real odors are evoked by odorant mixtures; thus, a gap will remain in understanding how the odors of mixtures relate to the combined molecular properties of their components. The simplest way to bridge this gap is to learn how odors of the mixtures relate to the odors of their components. To investigate these relations, odor qualities of vapors of 28 odorants, diluted to yield about the same odor intensities, and of their 168 binary, 112 ternary, and 28 quaternary mixtures were characterized using Harper's scale expanded from 44 to 136 descriptors. The odorants ranged from very unpleasant (isovaleric acid) to very pleasant (vanillian). The source levels for those 30 descriptors that were used most frequently were analyzed statistically. The scores for the mixtures tended to gravitate toward the arithmetic mean of the component scores. A simple linear regression equations was found for an approximate calculation of descriptor scores of mixtures from those of their components. Cases of enhancement of depression from this value, were observed.

Literature Cited

1. Amer. Soc. Test. Mater. E-544-75. *Standard Recommended Practice for Referencing Suprathreshold Odor Intensity*, ASTM, Philadelphia, PA (1975).

2. Dravnieks, A., Bock, F.C., Powers, J.J. Tibbetts, M., and Ford, M. *Comparison of Odors Directly and Through Profiling*, Chem. Senses & Flavor 3, 191-225 (1978).

3. Harper, R., Bate Smith, E.C., Land, E.C., and Griffitys, N.M. *A glossary of Odor Stimuli and Their Qualities*, Perf. Essent. Oil Rec. 59, 1-16 (1968).

4. Moskowitz, H.R., Dubose, C.N., and Reuben, M.J. *Flavor Chemical Mixtures--A Psychophysical Analysis*, in R.A. Scanlan, ed. Flavor Quality: Objective Measurement, pp. 24-44, ACS Symposium Series 51, Washington, DC (1977).

5. Dravnieks, A., and O'Neill, H.J. *Annoyance Potentials of Air Pollution Odors*. Am. Ind. Hyg. Assoc. J. 40, 85-95 (1979).

RECEIVED October 13, 1980.

The Efficacy of *n*-Aliphatic Alcohols and *n*-Aliphatic Fatty Acids on Various Membrane Systems with Special Reference to Olfaction and Taste

P. H. PUNTER and B. Ph. M. MENCO

State University Utrecht, Psychological Laboratory, Varkenmarkt 2, 3511 BZ Utrecht, The Netherlands

H. BOELENS

Naarden International, P.O. Box 2, 1400 CA Naarden, The Netherlands

1. Introduction

The nature of the receptor-sites responsible for odorous interactions has not yet been elucidated. Some experiments suggest the presence of specific proteinaceous receptors (1,2), whereas other data indicate the involvement of more non-specific lipidic or proteinaceous receptor moieties (3,4,5,6).

Homologous series of aliphatic n-alcohols and -fatty acids are useful to test the latter possibility, since numerous studies on membranes involve such compounds (e.g. 7,8). Previous studies using alcohols and fatty acids indicated that olfactory and gustatory thresholds for these compounds are closely related to chemotactic thresholds (4,5). The purpose of the present study is to expand these findings to other membrane-interaction systems, including numerous olfactory and gustatory threshold data supplied by various authors. Moreover, the implications of the present findings will be related to threshold measurements in general.

2. Procedure

There are several physico-chemical variables which need to be considered for the present study. These variables have been obtained as described in the following paragraphs.

2.1. <u>Saturated vapor pressures (SVP)</u>. All SVP's have been calculated using data given by Dreisbach (9). For both n-aliphatic alcohols and -fatty acids the log SVP is a linear function of the number of carbon atoms (N). For both functions the following regression equations have been obtained:

n-aliphatic alcohols: $\quad$ log SVP=-0.39 N-1.82 (r=0.99, t=25°)
$\qquad\qquad\qquad\qquad$ log SVP=-0.37 N-1.57 (r=0.99, t=37°)
n-aliphatic fatty acids: log SVP=-0.49 N-2.22 (r=0.99, t=25°)
$\qquad\qquad\qquad\qquad$ log SVP=-0.46 N-2.00 (r=0.99, t=37°)

in which $\underline{r}$ is the correlation coefficient and $\underline{t}$ the temperature in degrees celsius.

0097-6156/81/0148-0093$05.00/0
© 1981 American Chemical Society

2.2. <u>Solubility data</u>. Solubility data are taken from the literature (10,11,12). Solubilities can also been calculated from the octanol/water partition coefficient using the method of Hansch (13) or Yalkowsky and Morozowich (14). The following relationships have been found between the log solubility (S) and the number of carbon atoms (N) for the n-alcohols:

1. Bell (12) $\quad\quad$ log S=-0.58 N + 2.30 (t=25-30°)
2. Stephen and $\quad$ log S=-0.67 N + 2.68 (t=37°, C_4-C_7)
   Stephen (10)
3. Yalkowsky and $\quad$ log S=-0.59 N + 2.33 (t=30°, C_4-C_{12})
   Morozowich (14)
4. Hansch et al. $\quad$ log S=-0.61 N + 2.26 (t=15-25°, C_4-C_8)
   (13)

The correlation between log S (Mol/l) and the number of carbon atoms (N) is larger than 0.99 in all cases. For the n-fatty acids the following relationships have been found between log S and the number of C-atoms:

1. Bell (12) $\quad\quad$ log S=-0.60 N + 2.32 (t=25-30°)
2. Seidell (11) $\quad\quad$ log S=-0.65 N + 3.05 (t=20°, C_6-C_9)
3. Yalkowsky and $\quad$ log S=-0.61 N + 2.44 (t=20°, C_5-C_9)
   Morozowich (14)
4. Ralston and $\quad$ log S=-0.62 N + 2.77 (t=37°, C_5-C_{10})
   Hoerr (15)

As for the n-alcohols the correlation between log S and the number of carbon atoms is larger than 0.99 in all cases.

2.3. <u>The air/water partition coefficient (K^a/w)</u>. The air/water partition coefficient (K^a/w) can be calculated using the following formula (16):

$$K^a/w = \frac{\text{saturated vapor pressure } (^\circ K, \text{Mol/l}).}{\text{solubility } (^\circ K, \text{Mol/l}) \text{ in water}} \quad\quad (1)$$

Amoore and Buttery (17) suggest to use this formula only in cases in which the solubility in water at 25°C is smaller than 10 gram/l. For solubilities larger than 10 gram/l but not infinite they propose the following equation:

$$K^a/w = \left|\left|\left(\frac{55.5}{\text{sol.}}\right)- 0.0555 \mid M+1\right|P \times 0.97 \times 10^{-6}\right., \quad\quad (2)$$

in which sol. is the solibility in gram/l, P the SVP in mm Hg and M the molecular weight. For both n-fatty acids and n-alcohols the 25°C values of the air/water partition coefficients have been calculated using the solubility data from (12); for the 37°C values the solubilities given by (10) have been used for the n-alcohols

while for the n-fatty acids the solubilities from ($\underline{15}$) have been
used. Calculation of the linear regression between log K^a/w and
the number of carbon atoms (N) gives the following results for
the n-alcohols:

$$\text{at } 25^\circ C \quad \log K^a/w = -0.195 \text{ N} + 4.17 \text{ (r=0.99, } C_3-C_{12})$$
$$\text{at } 37^\circ C \quad \log K^a/w = -0.306 \text{ N} + 4.31 \text{ (r=0.99, } C_3-C_{12})$$

and for the n-fatty acids:

$$\text{at } 25^\circ C \quad \log K^a/w = -0.145 \text{ N} + 4.79 \text{ (r=0.97, } C_2-C_9)$$
$$\text{at } 37^\circ C \quad \log K^a/w = -0.190 \text{ N} + 4.99 \text{ (r=0.99, } C_2-C_9)$$

   2.4. Data treatment. Literature data on the efficacy of
n-alcohols and n-fatty acids in various model systems, organisms
and/or organs have been compiled and compared. The different
measures of efficacy used can be found in Tables 1 and 2 under
physiological or biophysical parameter. In the case of aqueous
solutions the log-efficacy was plotted against the number of
carbon atoms and linear regressions were calculated. In the case
of gaseous dilutions the concentration in air was corrected with
the air/water partition coefficient to the concentration in water
and subsequently the linear regression was calculated. If the
correlation between the log-efficacy and the number of carbon
atoms was significant to at least 5% the data were used for
further calculation. On basis of the slopes of the regression
lines the chemical potential ($\Delta\mu$) was calculated, assuming that
the chemicals are in equilibrium between the membrane and solution
phases. The following formula has been used ($\underline{4}$):

$$\Delta\mu(CH_2^O) = \alpha \times 2.3 \text{ RT } ^{cal}/\text{mole (1cal=0.239 J)},$$

in which α = the slope of the regression line of log-concentration
versus the number of C-atoms, R = the gas constant and T =
temperature in $^\circ K$.

3. Results

   Tables 1 and 2 present the relationship between the log-
efficacy and number of carbon atoms of the n-alcohols and n-fatty
acids for the different model systems investigated. For those
cases in which the range of compounds studied exceeded C_8 two
regression equations were computed. Table 3 presents the $\Delta\mu$ values
for the n-alcohols. The experiments cited have been classified in
four groups: anesthesia, chemotaxis, olfaction and taste. The
numbers refer to the data from Table 1. In order to investigate
whether there are significant differences between the mean $\Delta\mu$
values for the four different groups t-tests between the means
were computed. The results are presented in Table 4. Table 5
presents data analogous to Table 3 for the n-fatty acids.

Table 1. The Linear Regression Between the Log-Effectiveness (Physiological or Biophysical Parameter) and Number of Carbon Atoms for the _n_-Aliphatic Alcohols

	Model system or organism and/or organ	Detection method	Physiological or biophysical parameter	r[aa]	Range	Slope	Constant	Reference
1	Red blood cell ghost	Uptake	Anesthetic effect	-0.96	C5-C10	-0.60	1.30	(18)
				-0.99	C5-C8	-0.40	0.10	(18)
2	Red blood cells	Hemolysis	Inhibition of 50%	-0.99	C1-C10	-0.60	1.04	(19)
				-0.99	C1-C8	-0.58	0.99	(19)
3	Lobster axon	Electrophysiology	Anesthetic effect	-0.99	C1-C5	-0.59	1.37	(19)
4	Frog sciatic nerve	Electrophysiology	Anesthetic effect	-0.99	C1-C5	-0.43	0.61	(19)
5	Squid axon	Electrophysiology	Anesthetic effect	-0.96	C2-C8	-0.57	1.49	(19)
6	Tadpole	Reflex	Inhibition	-0.99	C2-C8	-0.56	0.61	(20)
7	Escherichia coli	Negative chemotaxis	Thresholds	-0.86[b]	C1-C4	-1.02	0.11	(21)
8	Physarum polycephalum	Chemotactic motive force	Thresholds	-0.99	C3-C10	-0.37	-0.56	(5)
9	Tetrahymena	Chemotaxis	Thresholds	-0.99	C1-C10	-0.41	-1.02	(5)
10	Nitella sp.	Chemotactic electrical response	Thresholds	-0.99	C3-C8	-0.64	0.87	(5)
11	Human olfactory organ[a]	Psychophysical response	Detection threshold	-0.84	C3-C12	-0.39	-2.68	(22)
				-0.97	C3-C8	-0.62	-1.52	(22)
12	Human olfactory organ	Psychophysical response	Detection threshold	-0.99	C3-C8	-0.55	-1.79	(22)
13	Human olfactory organ[a]	Psychophysical response	Detection threshold	-0.95	C3-C12	-0.36	-3.07	(23)
				-0.96	C3-C8	-0.49	-2.41	(23)
14	Human olfactory organ[a]	Psychophysical response	Detection threshold	-0.95	C1-C10	-0.86	-0.99	(16)
				-0.98	C3-C8	-0.86	-0.48	(16)
15	Rat olfactory organ	Behavioral response	Detection threshold	-0.92	C1-C12	-0.28	-1.10	(24)
				-0.93	C1-C8	-0.42	-0.59	(24)
16	Bat olfactory organ[a]	Indirect physiological methods	Detection threshold	-0.95[b]	C1-C4	-0.54	0.44	(25)
17	Human tongue	Psychophysical response	Taste threshold	-0.98	C2-C8	-0.49	0.49	(26)
18	Human tongue	Psychophysical response	Taste threshold	-0.97	C2-C10	-0.45	-4.90	(27)
				-0.98	C2-C7	-0.55	-4.49	(27)
19	Phormia regina tarsal taste hairs	Inhibition proboscis	Rejection threshold taste	-0.97	C1-C8	-0.65	1.76	(28)
20	Phormia regina	Behavioral response	Rejection threshold taste	-0.94	C1-C10	-0.65	1.48	(29)
				-0.94	C1-C8	-0.73	1.78	
21	Gryllus assimilis ovipositor	Tetanic vibratory response	Rejection threshold taste	-0.98	C1-C7	-0.85	3.18	(28)

[a]These threshold values have been measured in air and are corrected with the air/water partition coefficient to the concentration in water.

[aa]All r-values are significant at 1% except for those indicated with _b_, which are significant at 5%.

Table 2. The Linear Regression Between the Log-Effectiveness (Physiological or Biophysical Parameter) and the Number of Carbon Atoms for the *n*-Fatty Acids

	Model system or organism and/or organ	Detection method	Physiological or biophysical parameter	r[AA]	Range	Slope	Constant	Reference
22	Human erythrocyte	Anti hemolysis	Inhibition 50%	-0.87	C2-C18	-0.17	-2.12	(19)
23	Physarum Polycephalum	Chemotaxis	Threshold	-0.96	C3-C7	-0.14	-3.67	(5)
24	Nitella	Chemotaxis	Threshold	-0.99[a]	C4-C7	-0.42	-1.42	(4)
25	Human olfactory organ	Psychophysical response	Threshold normals	-0.66	C1-C10	-0.24	-2.50	(31)
26	Human olfactory organ	Psychophysical response	Threshold anosmics	-0.70	C1-C10	-0.23	-1.51	(31)
27	Human olfactory organ[A]	Psychophysical response	Threshold	-0.70	C2-C9	-0.19	-3.88	(23)
28	Human olfactory organ[A]	Psychophysical response	Threshold	-0.71	C2-C9	-0.31	-5.46	(22)
29	Human olfactory organ[A]	Psychophysical response	Threshold	-0.75	C2-C9	-0.36	-3.39	(30)
30	Human tongue[A]	Psychophysical response	Threshold	-0.91[a]	C2-C10	-0.14	-3.29	(27)
31	Dog olfactory organ	Behavioral response	Threshold	-0.81	C2-C8	-0.35	-10.25	(32)
32	Dog olfactory organ	Behavioral response	Threshold	-0.83[a]	C2-C8	-0.27	-3.06	(33)
33	Phormia regina, tarsal taste hairs	Inhibition proscobis	Threshold	-0.87	C2-C5	-0.15	-0.36	(34)

[A] These threshold values have been measured in air and are corrected with the air/water partition coefficient to the concentration in water.

[AA] All r-values are significant at 5% except for those indicated with a, which are significant at 1%

Table 3. The $\Delta\mu$ values for the n-aliphatic alcohols for four
different groupings, together with their means and
standard deviations. The data are taken from Table 1.

ANESTHESIA	CHEMOTAXIS	OLFACTION	TASTE
-536 5-8[*] 1[**]	-1367 1-4 7	-880 3-8 11	-696 2-8 17
-777 1-8 2	-496 3-10 8	-781 3-8 12	-781 2-7 18
-790 1-5 3	-549 3-10 9	-696 3-8 13	-871 1-8 19
-576 1-5 4	-857 3-8 10	-1221 3-8 14	-987 1-8 20
-764 2-8 5		-563 1-8 15	-1193 1-7 21
-750 2-8 6		-724 1-4 16	
$\bar{X}$ -699	-817	-811	-904
Sd. 112	404	226	193

[*] Range of carbon atoms in the regression equation on which the
$\Delta\mu$ values are based.
[**] This number refers to the serial number of the studies cited in
Table 1.

Table 4. t-Tests between the mean $\Delta\mu$ values of the n-aliphatic
alcohols for the four different groupings from Table 3.

	ANESTHESIA	CHEMOTAXIS	OLFACTION	TASTE
ANESTHESIA n=6				
CHEMOTAXIS n=4	t=0.70 df 8 n.s.			
OLFACTION n=6	t=1.09 df 10 n.s.	t=0.03 df 8 n.s.		
TASTE n=5	t=2.20 df 9 n.s.	t=0.43 df 7 n.s.	t=0.73 df 9 n.s.	

Table 5. The $\Delta\mu$ values for the n-fatty acids for four different groupings together with their means and standard deviations. The data are taken from Table 2.

ANESTHESIA	CHEMOTAXIS	OLFACTION	TASTE
-277 2-18[*] 22[**]	-563 4-7 24	-273 2-9 27	-199 2-10 30
	-190 3-7 23	-350 1-10 25	-201 2-5 33
		-327 1-10 26	
		-443 1-9 28	
		-511 2-9 29	
		-351 2-8 32	
		-497 2-8 31	
$\bar{X}$ -277	-376	-393	-200
Sd.		91	

[*] Range of carbon atoms in the regression equation on which the $\Delta\mu$ values are based.
[**] This number refers to the serial number of the studies cited in Table 1.

Since the number of experiments used is smaller than those for the n-alcohols it was not possible to do a statistical analysis. In Table 6 the mean $\Delta\mu$ values and the intercepts of the linear regression lines (from Tables 1 and 2) are compared for the n-alcohols and n-fatty acids.

4. Discussion and Conclusion

The n-alcohols and n-fatty acids can have different effects on a variety of biological functions associated with membranes. These effects can cause inhibition, stimulation or biphasic changes in membrane bound enzymatic systems ([7]). As can be seen from Tables 1 and 2, the efficacy of the n-alcohols and n-fatty acids is a linear function of chain-length: to obtain the same effect, a lower concentrations is needed as the chain-length increases. For the n-alcohols the increase in efficacy with increasing chain-length generally levels off for compounds with more than 8 carbon atoms. This effect is seen as a difference in the slope of the regression line of the whole range of alcohols tested and the slope of the regression line up to C_8.
According to Fourcans and Jain ([7]) and Jain and Wray ([35]) the crucial factor in the efficacy of alcohols to modify lipid

Table 6. The mean $\Delta\mu$ values and mean intercepts of the regression
lines of the n-aliphatic alcohols and n-fatty acids for
the four different groupings and overall. The data are
taken from Tables 3 and 5.

		ANESTHESIA	CHEMOTAXIS	OLFACTION	TASTE	OVERALL
N-ALCOHOLS	$\Delta\mu$	-699	-817	-811	-904	-802
	Sd.	112	404	226	193	230
	INT	0.86	-0.15	-1.06	0.54	0.04
	Sd.	0.52	0.82	1.04	2.97	1.68
	n	6	4	6	5	21
N-FATTY ACIDS	$\Delta\mu$	-277	-376	-393	-200	-344
	Sd.			91		132
	INT	-1.80	-2.54	-4.29	-1.83	-3.41
	Sd.			2.90		2.53
	n	1	2	7	2	12

structure and various functions (of membranes) is the hydrophobicity of the alcohol. Above a critical chain-length they cause
less perturbation in the lipid chains between which they are
intercalated, hence their efficacy is lower. For the n-fatty acids
it is difficult to find a similar effect; the reported ranges in
Table 2 are in most cases too small. As can be seen from the
correlation coefficients in Table 2 there is more scatter in the
fatty acid data than in the alcohol data. The correlation
coefficients are lower in most cases, although still
significant.
The results presented in Table 1 for the n-alcohols are all based
on interactions with lipid-protein systems. Results on lipid
systems only, show a similar trend. Table 7 summarizes a number
of these studies. The $\Delta\mu$ value for the data from Table 7 is
-858 cal/mole with a standard deviation of 221. This value is
very similar to the overall value for the lipid-protein systems
(Table 6).
In addition, dissociation constants based on electro-olfactograms

Table 7. The Linear Regression Between the Log-Effectiveness and Number of Carbon Atoms for the *n*-Aliphatic Alcohols on Different Model Systems

Model system	Detection method	Parameter	r[a]	Range	Slope	Constant	Reference
1. Dipalmitoyl-phosphatidylcholine	Fluorescence with chlorophyll as a probe	Concentration for a 5°C drop in the midpoint transition	-0.99	C4-C8	-0.55	1.28	(36)
2. Human erythrocyte phospholipid membranes	ESR with 3-spiro-cholestane as probe	Decrease ratio low versus midfield peaks (B/C ratio)	-0.99	C3-C8	-0.48	1.53	(20)
3. Black lipid membrane	Conductivity measurements	Resistance decrease	-0.98	C2-C7	-0.56	1.49	(20)
4. Water-mineral oil system	Interfacial tension	Concentration in aqueous phase to produce interfacial tension reduction of 1 dyne/cm^2	-0.99	C2-C6	-0.85	-0.99	(37)
5. Lipid monolayers of bovine olfactory organ	Surface tension	Increase of 1 dyne/cm^2	-0.99	C3-C8	-0.81	0.77	(38)

[a] All r-values are significant at 1%

for a series of n-aliphatic alcohols (6) showed that this parameter is in agreement with the findings presented in Table 1.
The $\Delta\mu$ values of the n-aliphatic alcohols based on the
dissociation constants are -1072 cal/mole for C_3 to C_8 and -871
cal/mole for C_3 to C_{10}.
The correlations between the dissociation constant and the number
of carbon atoms are -0.96 and -0.94 respectively. The same type of
linear relationship between effectiveness and chain-length has
been found for n-alkanes and n-thiols (39). Additional support for
the involvement of phospholipids in chemoreceptive processes can
be deduced from the fact that the thresholds for cis-aliphatic
compounds are similar or lower than those for trans-aliphatic
compounds (40; own unpublished results). This may be due to the
fact that aliphatic cis-compounds cause a greater disturbance in
the phospholipid bilayer than trans-aliphatic (41) compounds.
In the case of the n-alcohols the chemical potential ($\Delta\mu$) appears
to be quite similar (Table 3) in the biological systems which
have been examined here. This suggests that the nature of this
potential is a consistent property of membranes found in diverse
systems measured in a variety of ways. The t-tests over the means
for the four groupings (Table 4) do not show any significant
differences. In the case of the n-fatty acids (Table 5) it is
more difficult to make a meaningul comparison between the four
different groupings because of the limited amount of data.
Comparison of the $\Delta\mu$ values and intercepts of the regression lines
for the n-alcohols and n-fatty acids (Table 6) shows that the
behavior with regard to the effectiveness is rather independent of
the nature of the membrane system. The following conclusions can
be derived from Table 6:

1. From the $\Delta\mu$ values it can be deduced that the transfer from the
   water to the lipid phase takes more energy for the n-alcohols
   than for the n-fatty acids for the chain-lengths investigated.
2. From the intercepts it can be deduced that the sensitivity of
   the biological system is higher for the n-fatty acids than for
   the n-alcohols.

In the case of the n-alcohols the overall free energy of adsorption
($\Delta\mu$) is -800 cal/mole-CH_2. This value is in agreement with the
assumption that the process is controlled by hydrophobic interaction. According to Seeman (19) the hydrophobic region may
consist of:

a. non-polar portions of lipid molecules, and/or
b. non-polar interfaces between lipid and protein molecules, and/or
c. hydrophobic regions of protein molecules.

In the case of the n-fatty acids $\Delta\mu$ is considerably lower (Table 6).
Since the oil/water partition coefficients for these compounds are
not very different from those of the n-alcohols,it is suggested
that interactions of polar groups at the interface of the chemoreceptive membrane may be responsible for the difference (4).
Furthermore dissociation effects of these acids could play a role.
    This study points to the importance of hydrophobic membrane

regions in chemoreceptive processes. However, considerably disagreement remains about the actual role of these hydrophobic domains. The following opinions are quoted from a discussion in Hauser (42): "there is a clear possibility of phospholipids acting as a receptor", "it is hard to see how the interaction of a drug or a sweet molecule with a phospholipid can result in anything" and "couldn't it be possible that the phospholipids play a role in that they provide the micro-environment of the protein and that the motional state of the protein depends on this environment". This latter statement is in accordance with a conclusion from Fourcans and Jain (7) who state that many different membrane bound enzymes or enzyme systems from different sources exhibit partial or complete dependence upon membrane lipids for their activity.

It should be mentioned that data on cockroach antennal (43) and maxillar palp (44) olfactory sensilla show that different receptor cells display consistently different sensitivities towards the same ranges of n-alcohols (e.g. so called pentano and heptanol receptors). Additionally, the existence of vertebrate olfactory receptor cells which display different sensitivities for the same alcohols can be concluded from single-unit adaptation and cross adaptation studies (45). Although the effects could be due to different protein receptor species, they can also be explained on the basis of different lipid compositions in the receptor cells in question.

That rather specific proteins are also involved in chemoreceptive processes has been shown by several electrophysiological (1,2,45,46,47) and biochemical (48,49,50) studies. Moreover, freeze-fracture observations indicate the presence of a high intramembrane particle density in olfactory cilia when compared to non-sensory respiratory cilia (51,52,53). Therefore it is evident that membranes of olfactory sensory cilia differ from those of non-sensory kinocilia. Also microvilli from taste receptor cells display high intramembrane particle densities (54). From threshold measurements it can not be decided whether the hydrophobic domains act as sole receptor sites for the substances investigated, though this seems unlikely considering the above references. If so, the membranous particles could represent proteinaceous ion gates and/or transducting enzyme systems (e.g. membrane bound nucleotide cyclases) which are activated by the perturbation of the hydrophobic membrane domains. Alternatively these hydrophobic domains could act in conjunction with more specific proteinaceous receptor sites. In that case at least part of the intramembrane particles represent the actual receptor sites (53).

The studies cited in this paper show that for n-aliphatic alcohols (C_1-C_{12}) and -acids (C_2-C_9), olfaction and taste act in similar ways as chemotaxis and anesthesia. Jain et al. (55) came to a similar conclusion for many other membrane systems. Alcohols and fatty acids were used in the present study since olfactory and gustatory data on these compounds could be compared with those on many other systems. It should be kept in mind that threshold

determinations on other compounds may also only describe non-
specific interactions. Hence threshold determinations are of
limited value for answering specific mechanistic questions.
However, psychophysical studies could contribute by using
compounds with rather similar physical and physico-chemical
properties. Systematic quantitative threshold , self- and cross-
adaptation measurements using optical-, positional-, and cis-trans
isomers could provide useful data. Additionally, precise
assessments of olfactory and gustatory qualitative sensations may
provide more specific answers than quantitative assessments (see
e.g. 56,57,58). A combination of qualitative and quantitative
psychophysical experiments on the compounds suggested above could
be very useful especially in combination with electrophysiological
and biochemical studies.

List of abbrevations

$\bar{X}$ = mean values   Sd. = standard deviation of the mean
N = number of studies used   n.s. = not significant
df = degrees of
Int = intercept

Acknowledgments

    This study has been supported by grants from the Netherlands
Organization for the Advancement of Pure Research (15.31.06), the
National Institute for Water Supply (The Netherlands) and the
Centre National de Recherche Scientifique (Paris).

Summary

    The present study shows that n-aliphatic alcohols and fatty
acids have a similar efficacy for olfaction and taste as for
other membrane related detection systems, e.g. chemotaxis and
anesthesia. Using data of numerous authors,the change in chemical,
potential per CH_2-group added for the n-alcohols is -699 cal/mole
for anesthesia, -817 cal/mole for chemotaxis, -811 cal/mole for
olfaction, -904 cal/mole for taste with an average value of -802
cal/mole. For the n-aliphatic fatty acids these values are
respectively -277 cal/mole, -376 cal/mole, -369 cal/mole, -200
cal/mole and -330 cal/mole. The intercepts (in ^{10}log Mol/l) of
the regression lines of the efficacy versus chain-length for the
n-alcohols are 0.86 (anesthesia), -0.15 (chemotaxis), 1.04
(olfaction), 2.97 (taste) with an average value of 1.68. For the
n-aliphatic fatty acids these values are respectively -1.80, -2.54,
-5.07, -1.83 and -3.87. From these data it has been concluded that
irrespective of the membrane system,the transfer from the water to
the lipid phase takes more energy for the n-alcohols than for the
n-fatty acids (chemical potential values) and that the

investigated membrane-linked systems are more sensitive for n-fatty acids than for n-alcohols (intercepts). Suggestions for psychophysical experiments which may give more specific answers concerning the mechanisms of olfaction and taste are given.

Literature cited

1. Getchell, M.L. and Gesteland, R.C. The chemistry of olfactory reception: Stimulus specific protection from sulfhydryl reagent inhibition. Proc. Natl. Acad. Sci. USA, 1972, 69, 1494-1498.
2. Menevse, A.; Dodd, G. and Poynder, T.M. A chemical modification approach to the olfactory code. Studies with a thiol-specific reagent. Biochem. J., 1978, 176, 845-854.
3. Cherry, R.J.; Dodd, G. and Chapman, D. Small molecule lipid-membrane interactions and the puncturing theory of olfaction. Biochim. Biophys. Acta, 1970, 211, 409-416.
4. Ueda, T. and Kobatake, Y. Hydrophobicity of biosurfaces as shown by chemoreceptive thresholds in Tetrahymena, Physarum and Nitella. J. Membrane Biol., 1977, 34, 351-368.
5. Ueda, T. and Kobatake, Y. Changes in membrane potential, zeta potential and chemotaxis of Physarum polycephalum in response to n-alcohols, n-aldehydes and n-fatty acids. Cytobiology, 1977, 16, 16-26.
6. Senf, W.; Menco, B.Ph.M.; Punter, P.H. and Duyvesteyn, P. Determination of odour affinities based on the dose-response relationships of the frog's electro-olfactogram. Experientia, 1980, 36, 213-215.
7. Fourcans, B. and Jain, M.K. Role of phospholipids in transport and enzymic reactions. In: Advances in Lipid Research, 1974, Vol. 12, Paoletti, R. and Kritchevsky, D. (Eds.), Academic Press, New York, pp. 147-226.
8. Hansch, C. and Dunn III, W.J. Linear relationships between lipophilic character and biological activity of drugs. J. Pharm. Sci., 1972, 61, 1-19.
9. Dreisbach, R.R. Pressure-Volume-Temperature Relationships of Organic Compounds, 1952, Handbook Publ. Inc., 3rd ed., Ohio.
10. Stephen, H. and Stephen, T. Solubilities of Inorganic and Organic Compounds. Vol. 1, Binary Systems, 1963 (1979), Pergamon Press, Oxford.
11. Seidell, A. Solubilities of organic compounds, Vol. II, 1941, D. van Nostrand Company, New York.
12. Bell, G.H. Solubilities of normal aliphatic acids, alcohols and alkanes in water. Chem. Phys. Lip., 1973, 10, 1-10.
13. Hansch, C.; Quinlan, J.E. and Lawrence, G.L. The linear free-energy relationship between partition coefficients and the aqueous solubility organic liquids. J. Org. Chem., 1968, 33, 347-350.
14. Yalkowsky, S.H. and Morozowich, W. A physical chemical basis for the design of orally active prodrugs. In: Drug Design, 1980, Vol. 9, Ariëns, E.J. (Ed.), Acad. Press, New York. pp. 122-183.

15. Ralston, A.W. and Hoerr, C.W. The solubilities of the normal saturated fatty acids. J. Org. Chem., 1942, 7, 546-552.

16. Davies, J.T. and Taylor, F.H. The role of adsorption and molecular morphology in olfaction: The calculation of olfactory thresholds. Biol. Bull., 1959, 117, 222-238.

17. Amoore, J.E. and Buttery, R.G. Partition coefficients and comparative olfactometry. Chem. Sens. Flav., 1978, 3, 57-71.

18. Seeman, P.; Roth, S. and Schneider, H. The membrane concentrations of alcohol anesthetics. Biochim. Biophys. Acta, 1971, 225, 171-184.

19. Seeman, P. The membrane actions of anesthetics and tranquilizers. Pharm. Rev., 1972, 24, 583-655.

20. Paterson, S.J.; Butler, K.W.; Huang, P.; Labell, J.; Smith, I.C.P. and Schneider, H. The effects of alcohols on lipid bilayers: a spin label study. Biochim. Biophys. Acta, 1972, 266, 597-602.

21. Tso, W.W. and Adler, J. Negative chemotaxis in Escherichia coli. J. Bacteriol., 1974, 118, 560-576.

22. Van Gemert, L.J. and Nettenbreijer, A.H. Compilation of odour threshold values in air and water. Central Institute for Nutrition and Food Research, TNO, Zeist, The Netherlands, 1977.

23. Punter, P.H. Measurements of human olfactory thresholds for several groups of structurally related compounds. (in preparation)

24. Moulton, D.G. and Eayrs, J.T. Studies on olfactory acuity II. Relative detectability of n-aliphatic alcohols by the rat. Quart. J. Exp. Psychol., 1960, 12, 99-129.

25. Schmidt, U. Vergleichende Riechschwellenbestimmungen bei neotropischen Chiropteren. Z. Säugetierk., 1975, 40, 269-298.

26. Dethier, V.G. Taste sensitivity to homologous alcohols in oil. Fed. Proc., 1952, 11, 34.

27. Siek, T.J.; Albin, I.A.; Sather, L.A. and Lindsay, R.C. Comparison of flavor thresholds of aliphatic lactones with those of fatty acids, esters, aldehydes, alcohols and ketones. J. Dairy Sci., 1971, 54, 1-4.

28. Dethier, V.G. The limiting mechanism in tarsal chemoreception. J. Gen. Physiol., 1951, 35, 55-65.

29. Dethier, V.G. and Yost, M.T. Olfactory stimulation of blowflies by homologous alcohols. J. Gen. Physiol., 1951, 35, 823-839.

30. Patte, F.; Etcheto, M. and Laffort, P. Selected and standardized values of suprathreshold odour intensities for 110 substances. Chem. Sens. Flav., 1975, 1, 283-307.

31. Amoore, J.E.; Venstrom, D. and Davies, A.R. Measurement of specific anosmia. Percept. Motor Skills, 1968, 26, 134-164.

32. Neuhaus, W. Uber die Riechschärfe des Hundes für Fettsäuren. Z. verg. Physiol., 1953, 35, 527-553.

33. Moulton, D.G.; Ashton, E.H. and Eayrs, J.T. Studies in olfactory acuity IV. Relative detectability of n-aliphatic acids by the dog. Anim. Behav., 1960, 8, 117-128.

34. Chadwick, L.E. and Dethier, V.G. The relationship between chemical structure and the response of blowflies to tarsal stimulation by aliphatic acids. J. Gen. Physiol., 1947, 30, 255-262.

35. Jain, M.K. and Wray Jr., L.V. Partition coefficients of alkanols in lipid bilayer/water. Biochem. Pharmacol., 1978, 27, 1294-1295.

36. Lee, A.G. Interactions between anesthetics and lipid mixtures. Normal alcohols. Biochemistry, 1976, 15, 2448-2454.

37. Rathnamma, D.V. Mechanism of olfaction explained using interfacial tension measurements. In: Adsorption at Interfaces, Mittal, K.L. (Ed.), 1975, ACS symposium series, Vol. 8, The American Chemical Society. pp. 261-269.

38. Koyama, N. and Kurihara, K. Effect of odorants on lipid monolayers from bovine olfactory epithelium. Nature, 1972, 236, 402-404.

39. Patte, F. and Punter, P.H. Experimental assessment of human olfactory thresholds in air for some thiols and alkanes. Chem. Sens. Flav., 1979, 4, 351-354.

40. Kafka, W.A. Molekulare Wechselwirkungen bei der Erregung einzelner Riechzellen. Z. vergl. Physiol., 1970, 70, 105-143.

41. Pringle, M.J. and Miller, K.W. Structural isomers of tetradecenol discriminate between lipid fluidity and phase transition theories of anesthesia. Biochem. Biophys. Res. Comm., 1978, 192-198.

42. Hauser, H. Phospholipid model membranes: demonstration of a structure-activity relationship. In: Structure Activity Relationships in Chemoreception; Benz, G. (Ed.), IRL London, 1976. pp. 13-24.

43. Sass, H. Zur nervösen Codierung von Geruchsreizen bei: Periplaneta americana. J. comp. Physiol., 1976, 107, 49-65.

44. Altner, H.; Stetter, H. Olfactory input from the maxillary palps in the cockroach as compared with the antennal input. In: Olfaction and Taste, 7, Van der Starre, H. (Ed.), IRL London, 1980. In Press.

45. Baylin, F. and Moulton, D.G. Adaptation and cross-adaptation to odor stimulation of olfactory receptors in the tiger salamander. J. Gen. Physiol., 1979, 74, 37-55.

46. Revial, M.F.; Duchamp, A. and Holley, A. Odour discrimination by frog olfactory receptors: a second study. Chem. Sens. Flav., 1978, 3, 7-23.

47. Caprio, J. Olfaction and taste in the Channel catfish: an electrophysiological study of the response to amino acids and derivatives. J. comp. Physiol., 1978, 123, 357-371.

48. Koshland, D.E. Jr. Bacterial chemotaxis. In: The Bacteria, Vol. 7, Sokatch, J.R. and Ornston, L.N. (Eds.), Acad. Press, New York, 1979. pp. 111-164.

49. Goldberg, S.J.; Turpin, J. and Price, S. Anisole binding protein from olfactory epithelium: evidence for a role in transduction. Chem. Sens. Flav., 1979, 4, 207-215.

50. Cagan, R.H. and Zeiger, W.N. Biochemical studies in olfaction: Binding specificity of radioactively labeled stimuli to an isolated olfactory preparation from rainbow trout (Salmo gairdneri). Proc. Natl. Acad. Sci.USA, 1978,75, 4679-4683.
51. Kerjaschki, D. and Hörandner, H. The development of mouse olfactory vesicles and their cell contacts: A freeze-etching study. J. Ultrastruct. Res., 1976, 54, 420-444.
52. Menco, B.Ph.M.; Dodd, G.; Davey, M. and Bannister, L. Presence of membrane particles in freeze-etched bovine olfactory epithelia. Nature, 1976, 263, 597-599.
53. Menco, B.Ph.M. Qualitative and quantitative freeze-fracture studies on olfactory and nasal respiratory epithelia surfaces of frog, ox, rat and dog. II: Cell apices, cilia and microvilli. Cell Tissue Res., 1980. In Press.
54. Jahnke, K. and Baur, P. Freeze-fracture study of taste bud pores in the foliate papillae of the rabbit. Cell Tissue Res., 1979, 200, 245-256.
55. Jain, M.K.; Gleeson, J.; Upreti, A. and Upreti, G.C. Intrinsic perturbing ability of alkanols in lipid bilayers. Biochim. Biophys. Acta, 1978, 509, 1-7.
56. Schiffman, S.S. Physicochemical correlates of olfactory quality. Science, 1974, 185, 112-117.
57. Davis, R.G. Olfactory perceptual space models compared by quantitative methods. Chem. Sens. Flav., 1979, 4, 21-35.
58. Boelens, H. and Haring, H.G. Molecular structure and olfactive quality. In: Olfaction and Taste 7, Van der Starre, H. (Ed.), IRL London, 1980. In Press.

RECEIVED November 3, 1980.

Olfaction and the Common Chemical Sense

Similarities, Differences, and Interactions

WILLIAM S. CAIN

John B. Pierce Foundation Laboratory and Yale University, New Haven, CT 06519

All mucosal epithelium (e.g., mouth, throat, eyes, anus) possesses chemical sensitivity. In fact, all skin possesses such sensitivity beneath the epidermis (1). Except for the specialized and localized receptors of olfaction and taste, the chemoreceptive elements of mucosal tissue comprise free (unspecialized) nerve endings. In the respiratory tract, free endings of three cranial nerves (trigeminal, glossopharyngeal, vagus) play a chemoreceptive role. These register the "feel" of cigarette smoke during inhalation, the "bite" of chili pepper, the "burn" of ammonia, the coolness of menthol, and so on. Such experiences comprise sensations of the common chemical sense. They may lack the qualitative range and richness of odors or tastes, but can nonetheless add much to the enjoyment of eating, drinking, and smoking, and even of fresh air. Crisp, invigorating air often gains its sensory character from concentrations of ozone sufficient to trigger common chemical sensations.

The motivation for controlled common chemical stimulation varies markedly. Some persons crave hot spicy food, whereas others avoid even a hint of pungency (2). The difference may lie in personal criteria for what to deem painful or how to interpret pain. Even weak common chemical stimuli may eventually evoke pain, a reason why the "chemistry" of this modality has appealed to persons who study air pollution, warning agents, industrial contaminants, and agents for crowd control. Examples of common airborne substances with particular effectiveness include: sulfur dioxide, formaldehyde, acrolein, chlorine, automobile exhaust, sulfuric acid, acetic acid, ammonia, nitro-olefins, nitrogen dioxide, and cigarette smoke. Hundreds of other less common substances can also evoke intense pungency. Dixon and Needham (3), and subsequently Alarie (4), drew attention to three classes of potent irritants: 1) thiol alkylating agents characterized by a "positive halogen" atom (e.g., chloracetophenone, bromobenzyl-cyanide, amides of iodoacetate and acrylate), 2) dienophiles, which contain an ethylenic double bond polarized by electron withdrawing groups (e.g., acrolein, benzilidene malonitrile,

0097-6156/81/0148-0109$05.00/0
© 1981 American Chemical Society

o-chlorobenzylidene malonitrile, β-nitrostyrene), and 3) certain
organoarsenicals in which arsenic operates at the trivalent state
(e.g., diphenylaminochlorarsine, ethyl dichloroarsine, diphenyl-
cyanoarsine). These classes share an ability to react with SH
groups in protein receptor molecules. The higher the reactivity,
the stronger is the irritant. Another mode of interaction with a
receptor protein, specifically nucleophilic cleavage of S-S linkages, can help to account for the irritant potential of an additional group of substances with little in common otherwise, e.g.,
sulfur dioxide, chlorine, hydrides, hydroxides, and secondary
amines.

Although the categorization of substances into those that
interact with SH groups and those that can break S-S linkages
accommodates many potent irritants, it leaves out those thousands
of substances that can act as mild irritants. Indeed, virtually
any odorous substance, even if benign at concentrations most commonly encountered, can evoke pungency at high concentrations.
The mechanism for this action may involve interaction with protein receptors in free nerve endings or induction of changes in
membrane permeability through a disruption of the lipid bilayer.

Functional comparisons. Because various odorous substances
evoke noticeable pungency as well as odor, they offer the opportunity to study two perceptual systems at once. Certain rare
persons with unilateral resection of the trigeminal nerve (see
Figure 1) can offer particularly useful information regarding how
much of what we loosely call "odor magnitude" actually comes
about through activation of the common chemical sense in the
nose. Figure 2 depicts psychophysical (stimulus-response) functions for 1-butanol derived from the normal and deficient nostrils of neurectomized patients (5). This commonly used odorant
obviously appeals to both olfaction and the common chemical
sense. Absence of the trigeminal nerve accounted for the large
difference between the functions for the two nostrils. The results made it possible to pose the question: Would instructions
to a normal subject to tease odor magnitude from overall magnitude yield a picture similar to that seen with unilateral trigeminal resection? Figure 3 confirms that normal persons can
indeed seem trigeminally deafferented under appropriate instructions (6). This finding encouraged further explorations of one
modality seen against the backdrop of the other in normal subjects.

Certain neurophysiological experiments in the tortoise and
the rabbit implied that the trigeminal system behaved unlike the
olfactory system in certain important temporal properties (7).
For example, the trigeminal nerve response lagged markedly behind
that of the olfactory nerve. Such peripheral neural data raised
the question of whether reaction time to pungency would fall
markedly behind reaction time to odor in human subjects. As
Figure 4 reveals, it did. Because olfactory and trigeminal

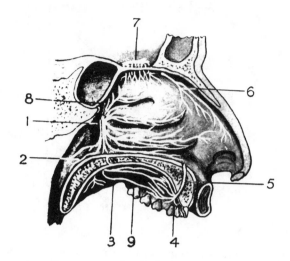

Figure 1.   Innervation of the lateral wall of the human nasal cavity (18).

The sphenopalatine ganglion (1) of the maxillary division of the trigeminal nerve gives rise to branches that mediate most common chemical sensations in the nose. Important branches include the posterior palatine nerve (2), the middle palatine nerve (3), the nasopalatine nerve (4, 5), posterior-superior lateral nasal nerve (8), and the anterior palatine nerve (9). The lateral nasal nerve (6) is derived from the ophthalmic division of the trigeminal nerve. The olfactory nerve (7) innervates only a relatively small portion of the cavity.

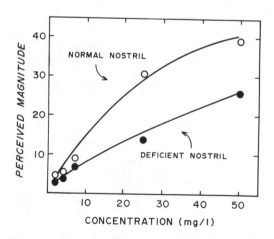

Figure 2.   Plot of how the perceived magnitude of 1-butanol varied with concentration in the normal (○) and the deficient (●) nostrils of patients with unilateral resection of the trigeminal nerve (data from Ref. 5)

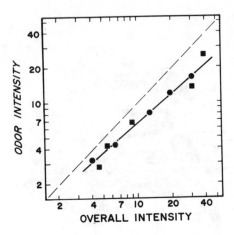

Figure 3.   Odor intensity vs. overall intensity of butanol in normal subjects (●). Also shown (■) is the perceived intensity (odor intensity) of butanol inhaled via the deficient nostrils of subjects with unilateral trigeminal destruction plotted against the intensity (overall intensity) of the stimulus inhaled via the normal nostrils of the neurectomized subjects (6).

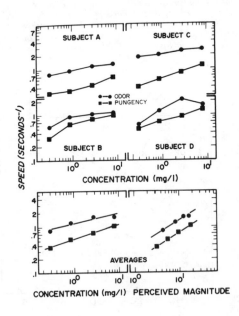

Figure 4.   (upper) Speed of response to odor and pungency of various concentrations of butanol. Data displayed for four subjects individually. (lower) Left side shows averages of the results in the upper portion plotted vs. concentration. Right side shows the same results plotted vs. the perceived magnitude of odor and pungency (6).

receptors occupy nearby loci, the temporal disparity could hardly arise from differences in transit time of the stimulus to the sensitive region. Furthermore, the difference (average: 890 ms) fell far outside the feasible range of differences in neural conduction time. Differences in the depth of the receptors seemed a more likely explanation (8).

Olfactory receptors contain long motile cilia. These distal structures, which apparently bear receptor sites, are covered with a layer of mucus. Approaching molecules must diffuse through this mucus. They must also diffuse through mucus to reach free nerve endings of the trigeminal nerve. In order to reach the nerve endings, however, the molecules must pass beneath the region of the respiratory or olfactory cilia and into intercellular spaces (Figure 5). This difference in the vertical component of molecular migration seems a reasonable account of the difference in latency between odor and irritation. A model that views diffusion through mucus as the rate limiting step in reception of airborne stimuli added a quantitative dimension to this conviction (9). When applied to the results on reaction time shown in Figure 4, the model estimated approximate equality of threshold concentration in the two modalities (Figure 6). Although we lack human data on the matter, neurophysiological data from the rabbit supports the conclusion (7). The model implied also that the receptors for pungency lie 110 μm below the air-mucus interface and that those for odor lie more superficially at 70 μm. These values seem like realistic approximations.

Depth and relative inaccessibility of receptor sites may also account for certain features of temporal integration in the common chemical sense. Tucker (7) noticed that the response of the trigeminal nerve of the rabbit increased from breath to breath during the first few breaths. The response of the olfactory nerve to the same stimuli, aliphatic alcohols, decreased or remained about the same. As in the case of response latency, a prediction that human beings would exhibit characteristics uncovered in the neurophysiological experiment actually held rather well. A psychophysical function for irritation (pungency) after three breaths fell above that for one breath (Figure 7). A function for odor after three breaths fell below that for one breath. For both olfactory and trigeminal stimulation, the molecules presumably remain in the mucus for at least a while after the termination of a sniff (10). The neural data suggest, however, that their presence has no influence on olfaction except during active flow through the nasal passages. The trigeminal nerve seems less dependent on flow for activation. Even between inhalations, the nerve exhibits some activity and such interbreath activity grows progressively. Conceivably, the sequestered locus of the free nerve endings causes a retarded rate of egress of molecules from the nerve endings as well as a retarded rate of progress toward the endings. Concentration may therefore build progressively breath by breath. Eventually, some process seems to limit

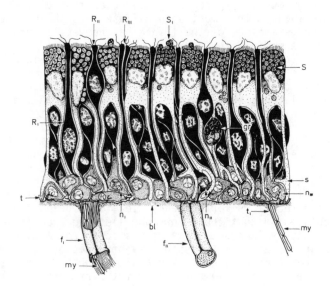

Figure 5.   Simplified diagram of the olfactory mucosa with the basic histological elements found in all vertebrates (19).

Receptor cells (R_1, R_{11}, R_{111}) are shown in black. Supporting cells (S) contain secretory granules and some protrusions in their free surface (S_1). Basal cells (n_1, n_{11}, n_{111}) are often neuroblasts differentiating into mature neurons. Note also granulocyte (gr). A basal lamina (bl) limits the deep part of the epithelium from the subadjacent lamina propria. Nerve fasicles (f_1, f_{11}) contain mainly olfactory axons, but also other myelinated (my) and unmyelinated (t, t_1) axons. The nonolfactory axons, often difficult to discern among the extensions of basal and supporting cells, generally have been thought to belong to the trigeminal nerve. The fibers seem to end among the basal and deep processs of supporting cells.

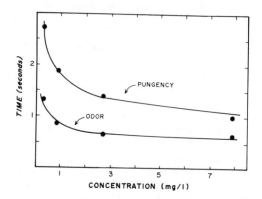

Figure 6.   Functions fitted to the latencies of pungency and odor of butanol.

These functions assume that net latency is derived from diffusion time to receptors plus irreducible reaction time. The small number of data points hardly provides a rigorous test of this recently elaborated model (9). The model has, however, already proved quite useful in descriptions of latency from single olfactory units, and merits thorough psychophysical testing.

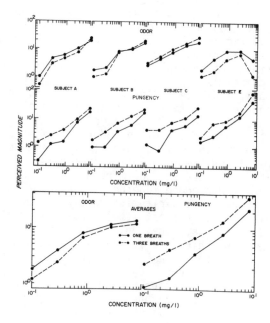

Figure 7.   Psychophysical functions for odor and pungency after exposures of one breath (——) or three breaths (– – –) (6). Upper portion depicts functions for individual subjects and lower portion depicts average functions.

further increases in response magnitude and adaptation (response decrement) ensues.

Interaction. In the study just described on the temporal integration of pungency (Figure 7), the data of one particularly sensitive subject implied that high levels of irritation might inhibit odor. The stability of the phenomenon in this person led to the question of whether strongly unbalanced stimuli (high in irritation, low in odor; high in odor, low in irritation) might uncover a general inhibitory interaction. The magnitude of the interaction actually proved far greater than anticipated (11).

As it turned out, some other investigators had previously noticed some interaction between olfaction and the common chemical sense. In a study of warning agents, Katz and Talbert (12) had remarked: "the odor of some irritants in higher concentrations is lost entirely in the pain of irritation in the nose." This remark describes an extreme and hence noticeable case. Discovery of the full range of possibilities requires experimental separation of odor and pungency. A substance like butanol, the stimulus used in the experiments shown in the previous figures, behaves like a mixture of odorant and irritant. Nevertheless, for butanol or for any other single stimulus, there exists no way to manipulate odor and irritation independently. This would require, in the ideal case, an actual mixture of odorless irritant and non-irritating odorant. Odorless irritants are difficult to find because virtually all irritants evoke odor. Carbon dioxide is one of a few major exceptions.

An experiment on possible olfactory-trigeminal interaction employed gas-phase mixtures of amyl butyrate, a fruity smelling odorant benign at moderate to low concentrations, and carbon dioxide, an odorless irritant at concentrations above 10%. Increasing amounts of carbon dioxide added increasing degrees of pungency to the fruity smell of the odorant. When asked to judge the degree of odor, pungency, or overall intensity of various concentrations of just amyl butyrate, just carbon dioxide, and mixtures of the two, subjects could render a picture of whether the sensory components added together linearly or whether they interacted. In the semilogarithmic coordinates of Figure 8A linear additivity would reflect itself in a family of parallel psychophysical functions. The converging trend, evident in the figure, reflects an inhibitory interaction. This becomes clearer in a view of how increasing amounts of carbon dioxide progressively inhibited odor (Figure 8C).

When plotted as a function of the concentration of amyl butyrate, the psychophysical functions take on a different character (Figure 9A). They still show convergence, but also remind us that odor intensity grows with concentration at a much lower rate than does pungency (6, 12). This gentle rate of growth shows up also in the inhibitory potential of odor upon irritation (Figure 9C).

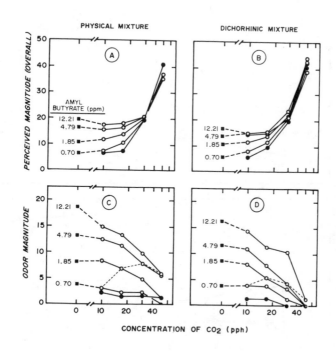

PHYSICAL MIXTURE   DICHORHINIC MIXTURE

CONCENTRATION OF CO$_2$ (pph)

Nature

Figure 8. (11).

(A) Perceived magnitude (linear scale) vs. concentration of carbon dioxide (logarithmic scale) for carbon dioxide presented alone (●), amyl butyrate presented alone (■), and mixtures of carbon dioxide and amyl butyrate (○). Parameter is concentration of amyl butyrate, indicated at left. Data points are medians taken across eight subjects.
(B) Same as A, but combinations of carbon dioxide and amyl butyrate presented dichorhinically, i.e., irritant (carbon dioxide) to one nostril and odorant (amyl buyrate) to the other. Data points are medians taken across ten subjects.
(C) Perceived odor component (denoted odor magnitude) of amyl butyrate alone (■), carbon dioxide alone (●), and physical mixtures (○). The low, but nonzero judgments for the odor of the odorless irritant carbon dioxide presumably reflect imperfect perceptual resolution between odor and irritation. The nonmonotonic function formed by the thin dashes depicts how odor magnitude would change in a case where concentration of odorant and irritant changed jointly. Compare this function with that of Subject E in Figure 7.
(D) Same as C, but dichorhinic mixtures.

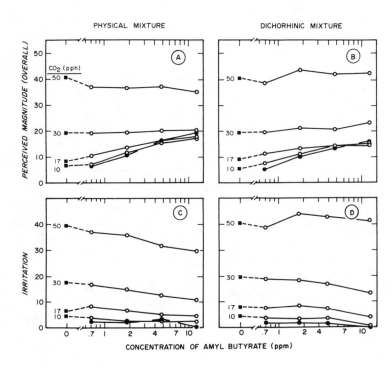

Nature

Figure 9.   (11).

(A) Same psychophysical data as in Figure 8A, but plotted here against concentration of amyl butyrate: amyl butyrate alone (●), carbon dioxide alone (■), and physical mixtures (○). Parameter is concentration of carbon dioxide, indicated at left.
(B) Same as A, but dichorhinic mixtures.
(C) Perceived irritating component of carbon dioxide alone (■), amyl butyrate alone (●), and physical mixtures (○).
(D) Same as C, but dichorhinic mixtures.

Dichorhinic mixtures, where one component enters one nostril
and the other component enters the other nostril, offered a useful way to discover whether the inhibitory interaction derived
from a fortuitous choice of stimuli that just happened to interfere with one another at the mucosa. If inhibition occurred in
the dichorhinic case, then it would establish two things: 1)
that the interaction depended less on a particular pair of odorant and irritant than on the activation of olfaction and the
common chemical sense by any suitable stimuli, and 2) that the
interaction probably took place in the brain. Figures 8B,D reveal that dichorhinic mixtures did indeed exhibit the interaction. Further experimentation indicated that this interaction,
almost indistinguishable from that seen in physical mixtures,
occurred in the brain (11).

Upon close inspection, it turns out that theneurophysiolo-
gical literature contains various indications that olfaction and
the common chemical sense would interact at a central locus.
For instance, Hughes and Mazurowski (13) and Sem-Jacobsen and
colleagues (14) noted that benign odorants will stimulate so-
called background activity in the olfactory bulb, whereas "sharp,
unpleasant odors" will inhibit it. It also turns out that the
inhibitory effect of carbon dioxide had been observed more than
a century before our experiments. Alexander Bain's textbook
The Senses and the Intellect (15) contains the tantalizing, yet
isolated, statement "If a current of carbonic acid accompanies
an odour, the effect [odor] is arrested."

Summary and Conclusions

The common chemical sense, particularly the portion mediated
by the trigeminal nerve, carries a considerable portion of the
chemosensory burden. It warns of the mere presence of highly
caustic substances and of high concentrations of almost all
organic agents. Its rather steep dose-response function, seen
as such both psychophysically (12) and neurophysiologically (16),
seems compatible with its role as a warning system. Aside from
this role, the common chemical sense adds an important and often
desirable dimension to chemosensory experience.

A degree of pungency or "feel" forms an intimate part of
many chemosensory experiences. Upon request, however, a person
can generally tease the common chemical attribute from an
olfactory-common chemical complex. It then becomes apparent
that olfaction and the common chemical sense obey somewhat different rules, particularly in the temporal realm. Common chemical sensations take somewhat longer to begin but last longer and
show more resistance to adaptation. These features may arise
from preneural events such as the time taken for molecules to
diffuse to free endings of the trigeminal nerve and from the
buildup of concentration in intercellular spaces in the epithelium. A diffusion-latency model of Getchell, Heck, DeSimone, and

Price (9) can serve to assess the effective depth of these endings and, perhaps, to assess the effective concentration (i.e.,
concentration at the membrane) necessary to evoke pungency.
Because almost all airborne organic substances possess some
ability to evoke such sensations, the mechanism of action of
mild irritants may be rather nonspecific and mere knowledge of
effective concentration at the neural membrane may account in
large measure for the nonuniformity in stimulating effectiveness.
Measures of latency (reaction time) will provide the appropriate
data to assess effective depth and effective concentration. In
addition, measures of the latency of common chemical sensations
can possibly serve a practical role in the assessment of the
chemosensory function of cigarette smokers (17) and persons exposed chronically to industrial contaminants (e.g., formaldehyde). These persons seem to develop tolerance to common chemical stimuli. What may appear adaptive may in fact represent such
nonadaptive changes as mucostasis and ciliastasis. A resulting
sluggish clearance of mucus would increase the effective depth
of the free nerve endings. Reaction time might therefore serve
as a nonsubjective indicator of chemosensory status.
   When seen over only a moderate range of perceived intensity
the olfactory and common chemical modalities may appear func-
tionally independent of one another. When seen over a wide range
and when stimulated with true odorant-irritant mixtures, the two
modalities show substantial interaction. Such interaction, apparently a central neural phenomenon, will generally serve to
give irritation sensory precedence over odor. The full conse-
quences of the phenomenon, like so many other interesting features of the rather poorly studied common chemical modality,
await further specification.

References

1. Keele, C. A. The common chemical sense and its receptors.
   Archives Internationales de Pharmacodynamie et de Thérapie,
   1962, 139, 547-557.
2. Rozin, P.; Gruss, L.; Berk, G. Reversal of innate aver-
   sions: attempts to induce a preference for chili peppers in
   rats. Journal of Comparative and Physiological Psychology,
   1979, 93, 1001-10014.
3. Dixon, M.; Needham, D. M. Biochemical research on chemical
   warfare agents. Nature, 1946, 158, 432-438.
4. Alarie, Y. Sensory irritation by airborne chemical agents.
   CRC Critical Reviews in Toxicology, 1973, 2, 299-363.
5. Cain, W. S. Contribution of the trigeminal nerve to perceived odor magnitude. Annals of the New York Academy of
   Sciences, 1974, 237, 28-34.
6. Cain, W. S. Olfaction and the common chemical sense: some
   psychophysical contrasts. Sensory Processes, 1976, 1,
   57-67.

7. Tucker, D. Olfactory, vomernasal and trigeminal receptor responses to odorants. In Y. Zotterman, Ed., "Olfaction and Taste"; Pergamon Press: Oxford, 1963; pp. 45-69.

8. Beidler, L. M. Comparison of gustatory receptors, olfactory receptors, and free nerve endings. Cold Spring Harbor Symposium on Quantitative Biology, 1965, 30, 191-200.

9. Getchell, T. V.; Heck, G. L.; DeSimone, J. A.; Price, S. The location of olfactory receptor sites: inferences from latency measurements. Biophysical Journal, 1980, 29, 397-412.

10. Hornung, D. E.; Mozell, M. M. Factors influencing the differential sorption of odorant molecules across the olfactory mucosa. Journal of General Physiology, 1977, 69, 343-361.

11. Cain, W. S.; Murphy, C. L. Interaction between chemoreceptive modalities of odour and irritation. Nature, 1980, 284, 255-257.

12. Katz, S. H.; Talbert, E. J. Intensities of odors and irritating effects of warning agents for inflammable and poisonous gases. Technical paper 480, Bureau of Mines, U. S. Department of Commerce, 1930.

13. Hughes, J. R.; Mazurowski, J. A. Studies on the supracallosal mesial cortex of unanesthetized, conscious mammals. II. Monkey. C. Frequency analysis of responses from the olfactory bulb. Electroencephalography and Clinical Neurophysiology, 1962, 14, 646-653.

14. Sem-Jacobsen, C. W.; Petersen, M. C.; Dodge, H. W., Jr.; Jacks, Q. D.; Lazarte, J. A.; Holman, C. B. Electric activity of the olfactory bulb in man. American Journal of the Medical Sciences, 1956, 232, 243-251.

15. Bain, A. "The Senses and the Intellect"; Longmans, Green: London, 1868.

16. Kulle, T. J.; Cooper, G. P. Effects of formaldehyde and ozone on the trigeminal nasal sensory system. Archives of Environmental Health, 1975, 30, 237-243.

17. Cain, W. S. Sensory attributes of cigarette smoking. Banbury Report No. 3, 1980, in press.

18. Turner, A. L., Ed. "Diseases of the Nose, Throat and Ear"; William Wood & Co.: Baltimore, 1936.

19. Graziadei, P. P. C.; Gagne, H. T. Extrinsic innervation of olfactory epithelium. Zeitschrift fur Zellforschung und Microskopische Anatomie, 1973, 138, 315-326.

Supported by grant ES 00592 from the National Institutes of Health.

RECEIVED October 20, 1980.

Odor and Molecular Vibration

Redundancy in the Olfactory Code

R. H. WRIGHT

6822 Blenheim Street, Vancouver V6N 1R7, Canada

The Dictionary defines redundancy as a condition of super-fluity or excess beyond the strict requirements of a given situation. In communication theory, unnecessary components of a signal are considered to convey no intelligence and are there-fore, regarded as useless or <u>redundant</u>. This may be true insofar as the transmission of a certain specific piece of information is concerned, but in fact, it is the presence of redundancy in person-to-person communications that makes possible the sort of richness and subtlety or expression without which life would lose much of its interest and meaning. Strictly speaking, exactly the same "information" is conveyed by the following two statements, yet in fact, they will evoke totally different responses in the reader:

1. The utmost parsimony in the quantitative deployment of any or all parts of speech is the incorporeal component inseparable from the apt expression and keen perception of those connections between ideas which awaken plea-sure and especially amusement.

2. Brevity is the soul of wit.

As an information channel with direct access into the conscious levels of the brain, the nose can recognize extremely subtle differences between odor sensations in a way which can only be achieved by the same sort of redundancy that gives spoken language its wealth of meaning.

- o -

The vibrational theory of odor postulates that the molecular attributes which confer olfactory specificity on each species of molecule are its low-frequency, "normal mode" oscillations ($\underline{1}$). The normal modes are the natural vibrational movements which can be excited independently of each other, and the low-frequency ones have frequencies corresponding to absorption in the far infrared with wave-lengths beyond 15-20 microns, or frequencies below about 750 cm-1. The objective evidence for this is derived from two kinds of experiment: the search for vibrational similarities in compounds whose odors are described as in some degree similar by qualified human observers, or by vibrational correlations with the reactions of insects to chemicals which elicit clustering or alarm responses or other indications that a "message" has been received and acted upon by the organism.

It is sometimes argued that insects are so remote from Man biologically that there can be little parallelism in their sensory physiology. As one of the most primitive of the senses, olfaction ranks with such basic mechanisms as nerve conduction, the genetic code, and the chiral specificities of organic molecules. To ignore the olfactory responses of insects would be like seeking to understand human genetics while ignoring Mendel's work with peas or the myriad studies of fruit flies.

- o -

Figure 1 shows how a specific olfactory response can be correlated with the vibrational attributes of the molecules which evoke that response. The low-frequency vibrations of a molecule are recorded by far infrared spectroscopy and are plotted as dots along a linear scale. The vibrational frequencies of other compounds which elicit the same response are added to give the "dot diagram" which shows a non-random distribution of frequencies with conspicuous clusterings at some places and gaps at others. To identify the statistically significant clusters or gaps, the number of dots in each 7 cm-1 interval is counted and plotted against the position of the interval so give the "Peak Number Plot" shown at the bottom. A line drawn through the plot is calculated from the formula,

$$P\nu = 10^{-3} \, M\nu 1/2d\nu$$

where $P\nu$ is the mean number of infrared absorption peaks to be expected in a randomly selected group of M compounds in a narrow wave number interval, $d\nu$, in the vicinity of frequency, ν. This is an empirical relation based upon some 500 spectra of a wide assortment of chemicals. Lines can be drawn two standard deviations above or below this to enable significant clusters or gaps in the dot diagram to be identified. This serves to identify, at least provisionally, the "favorable" and "adverse" elements in the vibrational pattern (2).

Figure 2 shows Peak Number Plots for several sets of compounds grouped on the basis of their ability to elicit a specific type of response in a particular species of insect. The validity of the resulting correlations has been verified experimentally by their predictive value. Thus, for example,

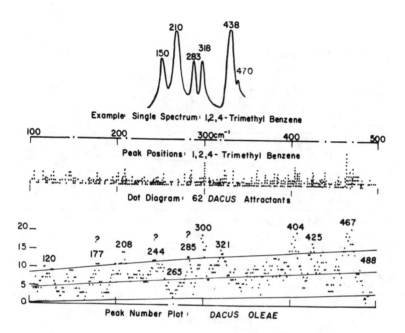

Figure 1.   Derivation of the peak number plot.

Peak frequencies of far infrared absorption spectra are plotted as dots along a linear scale. If the compounds have an odor in common or are specific attractants for a particular species of insect, the dots cluster at some places and avoid others. The number of dots in each 7-cm⁻¹ interval are counted and plotted to give the peak number plot. Lines drawn to standard deviations from the expected mean enable statistically significant favorable and adverse frequencies to be identified.

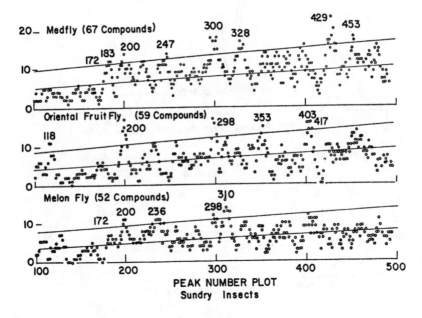

Figure 2. Peak number plots for the Mediterranean fruit fly, or medfly, the Oriental fruit fly, and the melon fly. There is some overlap in the patterns, and many compounds with frequencies corresponding to the overlaps will attract two, or in some cases, all three species of insect.

dl-homocysteine thiolactone hydrochloride was selected for test as an attractant for the olive fly, Dacus oleae, on the basis of its having frequencies at 212 and 463 cm-^{1}, with positive results while menthol, with adverse frequencies at 167 and 266 cm-^{1} acted as an "anti-attractant" for this insect when mixed with a standard lure (3). Again, when 2-heptanone was identified as the "alarm pheromone" of the ant, Iridomyrmex pruinosus, and a number of compounds mostly with chemically related structures were bioassayed, a Peak Number Plot enabled such totally dissimilar substances as triethylamine and heptyl butyrate to be tested and found to evoke the same "alarm response" as the natural pheromone (4). An even more unexpected result was obtained when the well-known biting-fly repellent, N,N-diethyl-m-toluamide, was shown to have vibrational similarities to several substances which attract the rhinoceros beetle, Oryctes rhinoceros, and was found to be distinctly alluring (5).

Such experiments based on insect responses have several advantages. For one, they have some potential economic significance in the control of insect pests without involving the environmental hazards associated with toxicants. Second, they often give statistically significant data based upon hundreds or even thousands of test subjects much more cheaply than could be got with human subjects. Finally, their responses are largely unequivocal: the insect either flies into a trap, or eats, or lays eggs, or it does not do these things. This contrasts sharply with human olfactory evaluations which are almost invariably hedged about with qualifications which, at times, make it difficult to determine how to classify an odorous stimulus. Thus, for example, the odor a m-ethyl nitrobenzene was described by expert perfumers as "weak almond with a trace of cumin alongside sassafras" (6).

Inspection of the Peak Number Plots shown in Figures 1 and 2 will make it plain that few chemicals will be likely to present the whole of the indicated frequency pattern to the organism's array of frequency-sensitive receptors. The pattern must, therefore, include enough redundancy to enable the organism to respond appropriately when only part of the pattern is presented. How small a part may depend upon how essential it is for the organism to recognize a precise message. Normally, this need for precision will be greater for sex-signals (pheromones) than for food or oviposition stimuli, so that "specialist receptors" are normally employed for pheromone reception, and "generalist receptors" for general purposes (7).

- o -

Somewhat unexpectedly, the Peak Number Plots have provided a new insight into the mechanism of olfactory stimulation, for it is evident that the frequency-elements they reveal do not relate to any specific stimulus but rather to the frequencies around which the stimulus frequencies must cluster if they are to evoke a response. In short, they are the frequencies to which the receptors are sensitive. On examination, it turns out that these receptor frequencies are spaced apart at equal frequency intervals, such that,

$$F = 12.8N - 6.4$$

where F is the frequency to which a given type of receptor is "tuned" and N is an integer. The possible light this throws on receptor mechanism has been considered elsewhere (8). For the present, it provides an impartial base for a given response-evoking pattern and for selecting candidate substances for test.

From Figure 2, the favorable frequencies for medfly attraction and the nearest evenly-spaced frequencies are shown in Table I.

Table I
Medfly Attractancy Pattern from the Peak Number Plot

Frequencies from the Plot	Nearest Frequency from the Formula
183	185.6
200	198.4
247	249.6
300	300.8
328	326.4
429	428.8
453	454.4
172 (adverse)	172.8

It seems clear that an insect like the medfly which is attracted to about 25% of a large and diverse selection of chemicals (9), must be able to respond to a relatively small sub-pattern drawn from the total pattern of medfly-favorable frequencies. The actual size of the minimum sub-pattern is suggested by the data summarized in Table II.

Table II
Redundancy in the Medfly Pattern

Number of Favorable Frequencies		Number of Compounds	Average Total Number of Peaks
0		3	4.7
1		13	5.1
2	out	14	7.5
3	of	18	7.8
4	7	8	8.6
5		7	11.0
6		2	13.5

It appears that a response is possible when there is no more than one needed element in the pattern coded into the molecule. In short, there is a very large element of redundancy in the pattern.

A second point suggested by the Table is the fact that where there are many favorable frequencies in the pattern, a randomly selected candidate is likely to have at least one that will approximately match one element of the pattern. This is no doubt the reason why 25.3% of the 2577 compounds tested by the U.S.D.A. attracted the medfly which has a seven element pattern. This contrasts with the fact that of 2618 compounds tested as attractants for the Mexican fruit fly, only 7.8% were effective (9). For this insect the Peak Number Plot shows only three favorable frequencies which makes it less probable that any given chemical will attract.

The manner in which insects show largely unqualified behavioral responses to odorous stimuli has in these and other ways provided a firm base from which to approach the matter of human responses to the same sort of signals.

- o -

Peak Number Plots for groups of compounds judged by expert human observers to have a fair degree of odorous similarity are shown in Figures 3, 4, and 5. Figure 5 includes the Plot for fifteen compounds whose odors are not related. These were compounds selected in 1966 by a Committee headed by Dr. L. M. Beidler as standard odor stimuli recommended for use in olfactory research (10). Their names and descriptions are shown in Table III. It is noticeable that in the absence of a common odor there is very little tendency to deviate from the expected mean frequency.

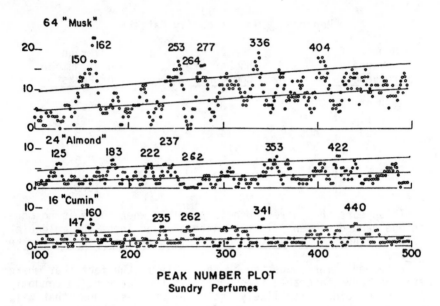

PEAK NUMBER PLOT
Sundry Perfumes

Figure 3.   Peak number plots for compounds having musky, bitter almond, and cumin odors

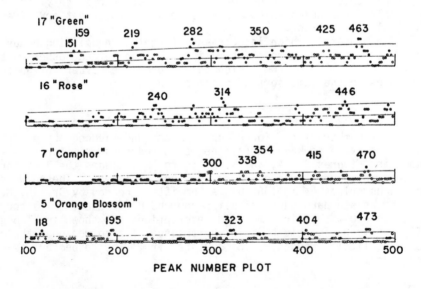

PEAK NUMBER PLOT

Figure 4.   Peak number plots for the green, rose, and orange blossom odors

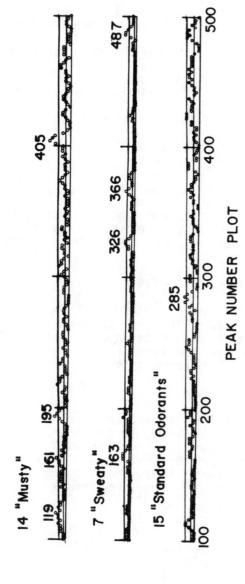

Figure 5.   Peak number plots for the musty and sweaty odors, and for 15 standard odorants which represent a variety of unrelated odors, thereby showing at most only one marginally significant deviation from the number of peaks to be expected in any randomly selected group of unrelated odorous compounds.

Table III
Recommended Olfactory Stimuli
(Odor Standards Committee, 1967)

Odor Class	Compound
Amber	Fixateur 404
Citrus	Methyl nonyl acetaldehyde
Camphor	Isoborneol
Floral	Dimethylbenzyl carbinyl acetate
	Indole
	Linalool
	Phenylethyl dimethyl carbinol
	Alpha terpineol
Fruity	p-Hydroxybenzyl acetone
Musky	Tonalid
	Musk 89
Sweet	Coumarin
Woody	Cedrol
	Thujamber

It is interesting and significant that the indicated favorable and adverse frequencies in these plots are clustered around the same set of evenly-spaced values as for insects, that is, those indicated by the formula,

$$F = 12.8N - 6.4$$

The even spacing is dramatically revealed by the "Peak Difference Plot" shown in Figure 6. This was developed by plotting differences between significant clusterings or gaps in 22 Peak Number Plots based on both human and insect evaluations, and counting the number of such differences in each 3 cm^{-1} interval.

- o -

Because human evaluations are nearly always hedged about with qualifications, there is usually some uncertainty in where the boundaries of a given odor class should be drawn. Thus, for example, six professional perfumers when asked to rate the degree of "green" and "rose" character in the compound "rose oxide" or 2-(2-methyl-1-propenyl)-4-methyl-tetrahydropyran, gave the following evaluations.

PEAK DIFFERENCE PLOT

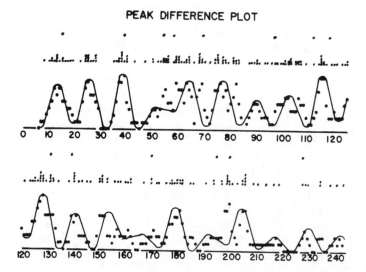

Figure 6.   The Peak difference plot is derived in the same way as the Peak number plot.

Note that differences between the positions of the peaks in the peak number plots are plotted as dots, and the number of dots appearing in each 3-cm⁻¹ interval are counted and plotted. The striking periodicity thereby revealed is interpreted as showing that the frequency sensitivities of the various biological receptors are evenly spaced.

Observer	Rose?	Green?	Comment?
1	"Weak"	"No"	"Geranium"
2	"Plus"		"Geranium"
3	"None"	"None"	"Naphthalenic"
4	"Strong"	"Strong"	
5	"Part"	"Part"	
6	"None"	"Medium"	"Stong peppery, weak flowery"

This rather mixed response is not unusual, and besides rendering it difficult to assemble satisfactorily large groups of compounds on which to base Peak Number Plots it doubtless underlies many of the conflicting claims of correlations between odor and one or another molecular attribute. Rose oxide was not one of the compounds used in constructing the Plot shown in Figure 4.

- o -

For present purposes it is probably better to use professional perfumers than untrained persons in assembling sets of compounds with similar odors. Not the least important is the experts' need to communicate one with another for which they must necessarily agree on their terminology. With untrained evaluators, the verbal descriptions will tell us more about the individual's background or associations. Thus, for example, a random selection of persons using the following words to describe the odor of methyl salicylate: "wintergreen", "peppermint", "chewing gum", and "liniment". The recent glossary of usage prepared by Harper et al. (11) is mainly in this latter category. Accordingly, the Peak Number Plots shown in Figures 3, 4, and 5 were based on perfumer-evaluations mostly communicated privately, but the Plot for the "musty" odor is based on data from Crocker and Dillon (12), and the "almond" and "cumin" patterns are from Klouwen and Ruys (6). The "sweaty" pattern is from evaluations by Amoore (13).

- o -

It is possible from these data to make a preliminary estimate of how much redundancy there is in human olfaction.

Taking 66 of the compounds used in constructing the Peak Number Plot for "musk" and rounding the observed peak positions to the nearest evenly-spaced value that was used in compiling Table II gives Table IV.

Table IV
Redundancy in the Musk Pattern

Number of Favorable Frequencies		Number of Compounds	Average Total Number of Peaks
0		5	6.2
1	out	12	6.1
2	of	19	6.5
3	6	22	8.3
4		8	9.5

As with the medfly pattern, in nearly every case where the number of "musk frequencies" was zero, there is at least one and usually several which, by a shift of a few units, would be close to an evenly-spaced value.   Furthermore, it must be recognized that the far infrared absorption spectra recorded with the samples dissolved in a solvent are subject to some displacement as a result of solvent effects which may vary somewhat when the solvent is changed.

A similar effect appears to operate when the stimulus moleule is in the near vicinity of a chiral receptor site (14). Evidently, the far infrared spectrophotometer is not the ideal piece of equipment for our purpose, and it has worked well enough to produce the various predictive successes the theory has so far achieved.

- o -

At this point the alert critic will point out that if as few as two elements of the musk pattern are sufficient for the musky sensation to be registered, then it should be possible for a great many compounds which have no musky odor to meet this reduced vibrational specification.   The correctness of this estimate is borne out by a scrutiny of the spectra of a random selection of 100 non-musk odorous or insect-attracting chemicals. The result is shown in Table V.

At first sight this looks like an unanswerable objection to the vibrational theory, and it is indeed unanswerable if it be held that the possession of one or two musk-favorable and no musk-adverse frequencies is a necessary and sufficient condition for the specific odorous sensation to be perceived.   But it has already been shown elsewhere that the probability of a quantum interaction and the stimulatory efficiency of a moleucle depend upon such additional factors as the frequency of the lowest vibrational mode which correlates with the threshold concentration (15), or the flexibility of the molecule which is related to the way the intensity of the sensation varies with the concentration of the stimulus (16). Also, it has been suggested that we perceive the musk sensation via "specialist receptors" which can be stimulated only by a molecule which can interact with more than one favorable frequency to match the plurality of sensitive sites the receptor deploys (8).

Table V
Vibrational Patterns of the 100 Non-Musks

Number of Musk-Favorable Frequencies	Number of Compounds	Number with Musk-Adverse Frequencies	Number of Potential Musk-Mimics
0	26	6	0
1	36	13	23
2	28	6	22
3	8	1	7
4	2	0	2

The probability of the same molecule making the necessary number of interactions to stimulate a single neuron will depend upon two things: the turn-around time between successive interactions, which must be short, and the diffusion rate, which must be small if the molecule is to stay near a given sensor long enough to make the necessary number of interactions. In this connection, it is interesting that all musky chemicals have relatively large molecules. Di-tert-butyl benzaldehyde (m.w. 218) is probably the simplest molecule with a musk odor (17). Typical musks, such as cyclohexadecanolide (m.w. 252) or musk xylol (m.w. 297) have distinctly higher molecular weights and correspondingly low diffusion rates. With only one exception (thujamber, m.w. 220), the one hundred compounds in Table V had molecular weights below 200 and most of them were below 160.

Evidently, then, a high molecular weight is also something that is necessary but not sufficient to evoke the musky odor. This illustrates an interesting aspect of the "scientific method".

- o -

The investigator searching for the specific molecular attributes which correlate with a specific odorous sensation, begins by assembling a group of compounds which have a common characteristic odor. He then looks for a common physical or chemical attribute, and, being human, when he finds one he is tempted to call it THE crucial attribute. It follows that the various "competing" olfactory theories - vibrational, structural (18) or stereochemical (19) - are not so much alternatives as complements. Each fills a gap in the other's picture.

Thus, given that the initial act in a stimulus-receptor interaction is the transfer of a quantum of vibrational energy from an excited receptor to an unexcited stimulus (8), the probability of the transfer would depend upon the relative orientations of the two vibrating dipoles, which would, in turn, depend upon which of its several profiles the stimulus presents and also upon the "shape" of the physical oscillation which has to take the energy quantum away from the receptor.

In short, the vibration patterns and the molecular profile correlations are complementary factors in determining the specific signal the stimulus passes into the organism.

- o -

A similar comparison can be made to establish the approximate amount of redundancy in the "bitter almond" stimulus pattern. Table VI shows a clear evidence of redundancy.

Table VI
Redundancy in the Bitter Almond Pattern

Number of Favorable Frequencies	Number of Compounds	Average Total Number of Peaks
0	1	4
1 out	8	6.9
2 of	7	6.9
3 6	6	6.7
4	0	-
5	2	8.5

The fact that the amount of redundancy seems to be about the same as for musk is interesting in view of the possibility that one sensation is being received through specialist and the other through generalist receptors (7).

Once again, it can be pointed out that with the indicated amount of redundancy, there should be many compounds able and willing to present one or two almond-favorable and no almond-adverse frequencies. Taking the same 100 compounds that were used in compiling Table V, and making a similar break-down with respect to the bitter almond pattern, we have Table VII.

Table VII
Vibrational Patterns of 100 Assorted Chemicals

Number of Almond-Favorable Frequencies	Number of Compounds	Number with Almond-Adverse Frequencies	Number of Potential Almonds	Number of Actual Almonds
0	21	4	17	0
1	43	11	32	3
2	20	7	13	2
3	13	3	10	4
4	1	0	1	
5	2	0	2	2

Evidently, again, there is evidence of a necessary-but-not-sufficient correlation, but the case is, perhaps, different from that of musk. The odor of musk is distinct even when blended with or accompanied by other "notes" such as amber or jasmin. The bitter almond odor is more easily submerged and "lost" in a complex blend of odors so that the only way to isolate it clearly is to fatigue the nose with respect to one part of the pattern and then "look for" the residual parts. In this way, the "community of odor property" in two sensations can be estimated in most of the sensations received via the generalist receptors (20).

- o -

For the full story of stimulus specificity to be told, many things remain to be worked out. Selective fatiguing will help to sort out the various notes in the overall sensation, while, on the physical side the various far infrared absorption frequencies must be given "vibrational assignments". An assignment is an unambiguous picture of the way the molecular shape changes during the oscillation. To take a simple example, chlorobenzene, there is a "wagging mode" when the chlorine swings from side to side with respect to the benzene ring. There is also a "stretching mode" that can be pictured as resulting from opposite ends of the molecule being pulled apart and then let go. Yet again, there is a "breathing mode" in which the benzene ring swells and shrinks, and so on. It is a matter of great difficulty to sort out and identify all the modes of even so simple a molecule as chlorobenzene, so that it is likely to be a long time before the vibrational assignments of a compound like musk xylol or cyclopentadecanone can be made known.

For the time being, then, it will be necessary to work in the half-light which is the best that existing knowledge can throw on our problem. That this can still provide some insights into the olfactory complexities that confront us is, perhaps, best illustrated by a couple of examples.

The many qualifications made by expert "noses" were referred to above as constituting an obstacle to finding Peak Number Plots for particular odors.   However, once a few such Plots have been developed it then becomes possible to take a fresh look at the comments and qualifications and begin to trace their origins in the respective molecular species.

Thus, for example, the expert evaluations of "rose oxide" mentioned rose, green and also a naphthalenic note. The compound has peaks at 284 and 425 which are close to two in the green pattern (Figure 4), and one at 314 matching one in the rose pattern, and its peaks at 172, 372 and 470 are fairly near the three far infrared peaks of naphthalene (in benzene), namely 180, 360 and 475 cm-[1].

A second example is provided by perilla aldehyde whose odor has been described as including green, cumin and almond notes. Its far infrared spectrum has peaks at 156 and 234 closely matching ones in the cumin pattern; its peaks at 234 and 420 are near those at 237 and 422 in the almond pattern, and its peaks at 156, 277 and 420 are not far from those at 159, 282, and 425 in the green pattern.

The relative weights to be given to the various notes making up the overall sensation will depend only partly upon the objective attributes of the stimuli (16).   What may sometimes be more important is the subjection factor: what the observer is "looking for" in the sensation.

Given the available evidence, it would appear that where human evaluations are concerned there is enough redundancy in the mechanism for a certain type of sensation, rose, green, almond, or whatever, to be recognized given at least one but more usually at least two elements of the pattern in a stimulus.

With a wider range and number of far infrared absorption spectra and more of the requisite expert evaluations, it would doubtless be possible to extend and eventually to clarify our understanding of the complexities of human odor evaluations. The existence of these complexities is at once an obstacle and a challenge and is a consequence of the fact that the messages coming into our consciousness are complex because the molecules in which they are conveyed are themselves complex. What is important is the fact that the sensory inputs can be put in a one-to-one relation with the molecular-vibrational attributes of the stimuli that induce them.

It is to be hoped that somehow and somewhere, means will be found to finance and carry out the systematic compilation of high quality far infrared absorption spectra of odorous compounds together with a systematic evaluation of the odors associated with them.

LITERATURE CITED

1. Wright, R.H. "Odor and Chemical Constitution." Nature, 1954, 173, 831.

2. Wright, R.H. and Burgess, R.E., "Molecular Coding of Olfactory Specificity." Can. J. Zool., 1975, 53, 1247-1253.

3. Wright, R.H., Chambers, D.L. and Keiser, I. "Insect Attractants, Anti-Attractants and Repellents." Can. Ent., 1971, 103, 627-630. See also, Stavrakis, G.N. and Wright, R.H., "Molecular Vibration and Insect Attraction: Dacus oleae." Can. Ent., 1974, 106, 333-335.

4. Wright, R.H. and Brand, J.M., "Correlation of Ant Alarm Pheromone Activity with Molecular Vibration." Nature, 1972, 239, 225-226.

5. Madison, P.A. personal communication.

6. Klouwen, M.H. and Ruys, A.H., "Chemical Constitution and Odor. III. Comparison of the Odor of Benzaldehyde and Nitrobenzene Derivatives with the Same Substituted Groups." Parfeum. Kosmetik, 1962, 43, 289-292. Also in Parfum. Cosmet. Savons, 1963, 6, 6-12.

7. Wright, R.H. "Specific Anosmia: A Clue to the Olfactory Code or to Something Much More Important?" Chem. Senses and Flavor, 1978, 3, 235-239.

8. Wright, R.H. "Odor and Molecular Vibration: Neural Coding of Olfactory Information." J. Theor. Biol., 1977, 64, 473-502.

9. Beroza, M. and Green, N. "Materials Tested as Insect Attractants." U.S. Department of Agriculture Handbook No. 239, Washington, D.C., June, 1963.

10. Wright, R.H. "Far Infrared Spectra of Some Odorous Compounds." Am. Perfumer and Cosmetics, 1968, 83, 43-45.

11. Harper, R., Land, D.G. and Griffiths, N.M. "Odor Qualities: A Glossary of Usage." Br. J. Psychol., 1968, 59, 231-252.

12. Crocker, E.C. and Dillon, F.N. "Odor Directory." Am. Perfumer, Essent. Oil Rev. 1949, 53, 297-301, 396-400.

13.   Amoore, J.E., Venstrom, D. and Nutting, M-D., "Sweaty Odor
      in Fatty Acids: Measurement of Similarity, Confusion and
      Fatigue." J. Food Sci., 1972, 37, 33-35.

14.   Wright, R.H. "Odor and Molecular Vibration: Optical Iso-
      mers." Chem. Senses and Flavor, 1978, 3, 35-37.

15.   Wright, R.H. "Odor and Molecular Vibration: A Possible
      Membrane Interaction Mechanism."   Chem. Senses and Flavor,
      1976, 2, 203-206.

16.   Wright, R.H. "The Perception of Odor Intensity: Physics or
      Psychophysics."   Chem. Senses and Flavor, 1978, 3, 73-79,
      241-245.

17.   Beets, M.G.J. "Structure and Odor." Soc. Chem.   Ind. Mono-
      graph No. 1.   London, 1957, 54-90.

18.   Beets, M.G.J. "Structure-Activity Relationships in Human
      Chemoreception."   Applied Science Publishers Ltd. Barking,
      Essex, 1978.

19.   Amoore, J.E.   "Molecular Basis of Odor."   C.C. Thomas,
      Springfield, 1970.

20.   Cheesman, G.H. and Mayne, S. "The Influence of Adaption on
      Absolute Threshold Measurements for Olfactory Stimuli."
      Quart. J. Exp. Psychol., 1953, 5, 22-30.

RECEIVED October 30, 1980.

Computer-Assisted Studies of Chemical Structure and Olfactory Quality Using Pattern Recognition Techniques

PETER C. JURS and CHERYL L. HAM

Department of Chemistry, Pennsylvania State University, University Park, PA 16802

WILLIAM E. BRÜGGER

International Flavors and Fragrances, Inc., 1515 Highway 36, Union Beach, NJ 07735

The attempt to rationalize the connection between the molecular structures of organic compounds and their biological activities comprises the field of structure-activity relations (SAR) studies. Correlations between molecular structure and biological activity are important for the development of pharmacological agents, herbicides, pesticides, chemical communicants (olfactory and gustatory stimulants) and for the investigation of chemical and genetic toxicity. Practical importance attaches to these studies because the results can be used to predict the activity of untested compounds, e.g., design drugs. In addition SAR studies can direct the researcher's attention to molecular features that correlate highly with biological activity, thus confirming or suggesting mechanisms or further experiments. SAR studies have been used to some extent in the pharmaceutical and agricultural industries. The methods are beginning to be applied to the important problems of chemical toxicity and chemical mutagenesis and carcinogenesis.

The superior way to develop predictive capability is to understand, at the molecular level, the mechanisms that lead to the biological activity of interest. Unfortunately, this knowledge is not yet available for most classes of biologically active compounds. Furthermore, the progress made through a living system by an active compound or its precursors is not usually known. Thus, two choices are presented: study the mechanisms for a very few compounds to develop fundamental information for those few compounds, or use empirical methods to study larger sets of compounds with correlative methods. The latter method comprises an SAR approach to the problem. Thus, one has available a set of compounds that have been tested in a standard bioassay and the observations that resulted from the tests. One can then search for correlations between the structures of the compounds tested and the biological observations reported. One is actually modelling the entire

0097-6156/81/0148-0143$05.00/0

process of uptake, transport, distribution, metabolism, cell
penetration, receptor binding, excretion, etc.

The discovery and design of biologically active compounds
(drug design) is a field that has been subject to widespread
and well-documented (1-8) changes in the past decade. A host
of new techniques and perspectives has evolved. While these
techniques have been used largely for the development of pharma-
ceuticals, they can also be applied to the rationalization of
structure-activity relations among sets of toxic, mutagenic, or
carcinogenic compounds and to studies of olfactory stimulants.

Several approaches to SAR have been reported: the semi-
empirical linear free enrgy (LFER) or extrathermodynamic model
proposed by Hansch and coworkers (9,10,11), the additivity or
Free-Wilson model (12); quantum mechanically based models (13,
14) and pattern recognition methods (8,15). Reviews are cited
that describe the progress made using each of the approaches.

Structure-Activity Studies of Olfactory Stimulants

Several theories relating molecular properties to perceived
odor quality have been advanced. Examples include the work of
Wright (16,17) who links odor quality to molecular vibrations
in the far-infrared, and of Amoore (18) who links odor quality
to molecular shape, size, and electronic nature and who introduced the concept of primary class. Beets (19) has discussed
odor quality relative to molecular shape as represented by
oriented profiles, chirality, and functional groups. In a
recently published book (20) he has expanded these discussions.
Theimer and coworkers (21,22,23) have discussed the importance
of the molecular cross-sectional areas, free energies of de-
sorption, and chirality in relation to odor. A discussion of
musk odor quality and molecular structure has been presented
by Teranishi (24). Laffort and coworkers (25) have related
odor quality to four molecular properties derived from gas
chromatographic retention indices measured on four stationary
phases.

Focussing on a few molecular parameters at a time does not
allow predictions of odor quality for large collections of
compounds. Studies have appeared in which diverse sets of
molecular parameters have been investigated simultaneously using
methods that can handle many parameters at once, e.g., multiple
linear regression analysis. Schiffman (26) used multidimensional
scaling techniques to study correlations between 25 physicochemical parameters and the olfactory qualities of 39 odorants.
The physicochemical parameters used included molecular size,
weight, number of double bonds, functional groups, solubility,
and Raman spectral bands. Another study (27) expanded the
work to 19 different compounds and generated similar conclusions.
Dravneiks (28) used 14 structural features and multiple linear
regression analysis to find linear equations that fit measured

intensity, threshold, and odor quality data. Dravneiks (29) used molecular weight, 38 attributes derived from Wiswesser Line Notation representations of molecular structures, and combinations of these parameters (118 indices in all) to seek correlations with odor intensities and vapor pressure of olfactory stimulants. Boelens (30) used multiple linear regression analysis of physicochemical parameters to study a set of compounds with musk and bitter almond odors. The 1-octanol/water partition coefficients, gas chromatographic retention indices, and molecular shape and volume parameters of the odorants (4 parameters total) were used. He obtained equations for 16 bitter almond compounds and for 16 musk compounds relating the four parameters to odor quality with multiple correlation coefficients of 0.95 and 0.93. Greenberg (31) found strong correlations between the 1-octanol/water partition coefficient of odorants and their intensities using multiple linear regression analysis. McGill and Kowalski (32) used pattern recognition methods to investigate relationships between molecular structure and odor quality. The electron donor ability and directed dipole of compounds were found to be related to odor quality. Brügger and Jurs (33) used pattern recognition methods to identify 13 calculated molecular structure descriptors that could classify odorants as musks or nonmusks. A data set of 240 nonmusks and 60 musks was used to derive the classifier. The classifier was used to predict the odor quality of nine unknown compounds, and all were classified correctly as musk odorants.

Methodology for SAR Studies

The fundamental premises involved in applying pattern recognition methods to SAR studies are as follows.

- Molecular structure and biological activity (olfactory quality) are related.

- The structures of compounds having a particular odor quality and compounds of similar structural classes that do not can be adequately represented by a set of molecular structure descriptors.

- A relation can be discovered between the structure and activity by applying statistical and pattern recognition methods to a set of tested compounds.

- The relation can be extrapolated to untested compounds.

The heart of the approach is finding a set of adequate descriptors for a particular data set consideration, that is, a set of descriptors for which a discriminating relation can be found.

The structure-activity studies described here involve the ADAPT (automatic data analysis using pattern recognition techniques) computer software system. This system has been developed

over the period from 1974 to the present. It is fully operational and has been reported in the scientific literature (8,
34-36). Research performed on the ADAPT system has also been
reported in a number of publications (33,37,42).

The ADAPT system currently consists of approximately sixty
programs written in the FORTRAN language and meant to be
executed interactively on a minicomputer or a larger time-
sharing computer. Development at Penn State has been on a
MODCOMP II/25 16-bit minicomputer with 65,000 16-bit words of
core memory. The system has been designed and implemented to
provide the user with all the capabilities necessary to perform
SAR studies on sets of up to several hundred compounds at a
time.

The fundamental steps involved in performing an SAR study
using this system are shown in Figure 1. The individual steps
are as follows:

(a) Identify, assemble, input, store, and describe a data
set of structures for chemicals that have been tested for biological activity.

(b) Develop computer generated molecular descriptors for
each of the members of the data set. The descriptors may be
derived directly from the stored topological representations
of the structures, or they may require the development of three
dimensional molecular models.

(c) Using pattern recognition methods, develop classifiers
to discriminate between active and inactive compounds based on
the sets of molecular descriptors.

(d) Test the predictive ability of these discriminants on
compounds of unknown activity.

(e) Systematically reduce the set of molecular structure
descriptors employed to the minimum set sufficient to retain
discrimination between the active and inactive compounds and
to retain high predictive ability.

Entry of Molecular Structures. The ADAPT system has as
one of its components all the modules necessary to enter,
modify, retrieve, and draw molecular structures of organic
molecules. This portion of ADAPT has been operational for
several years and has been employed in several published studies.
The routines allow the convenient, interactive entry of structures by sketching them on the screen of a graphics display
terminal. This can be done in thirty seconds to several minutes
per compound, depending on structural complexity. No special
techniques beyond those used in sketching molecular structures
on a blackboard are needed. Thus, structure files on the order
of hundreds of compounds can be entered into ADAPT in reasonable
amounts of time. The structure files are stored permanently on
disc files for further processing by the other modules of ADAPT.
Information saved for each compound includes a compressed
connection table, ring information, a list of associated numerical

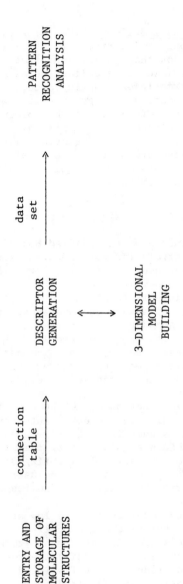

Figure 1. Flow chart of steps involved in structure–activity studies using chemical structure information, handling and pattern recognition methods

information, an identification number, the chemical name of the
compound, and the two-dimensional coordinates of the atoms as
entered (for possible redrawing later or for starting coordinates
for modelling).

   Molecular Mechanics Model Builder. The three-dimensional
molecular model builder routine interfaced to ADAPT (MOLMEC) is
used to derive information on the spacial conformation of molecules. A molecule can be viewed as a collection of particles
held together by simple harmonic or elastic forces. These
forces can be defined by potential energy functions whose terms
are functions of the atomic coordinates of the molecule. This
function can then be minimized to obtain a strain-free three-
dimensional model of the molecule. In the strain minimization
section, the atom coordinates are systematically altered until
a minimun is found in the strain or potential energy function.
The strain function used in MOLMEC is:

$$E_{strain} = E_{bond} + E_{angle} + E_{torsion} + E_{non-bond} + E_{stereo}$$

   The bond and angle functions are modified Hooke's Law
functions. The torsional strain for carbon-carbon single bonds
is a function containing the usual $(1 + \cos 3\theta)$ term but para-
meterized to provide the known values for butane. The nonbonded
strain term is an exponential-six function. The last term of
the function has been added to assure the proper stereochemistry
about an assymetric atom. An adaptive pattern search routine
is used to minimize the strain energy because it does not require analytical derivatives. The amount of time necessary to
obtain good molecular models depends upon the number of atoms in
the molecule, the initial strain of the molecule, and the degrees
of freedom in the structure.
   The graphics interaction section of MOLMEC contains routines
capable of rotating and aligning the molecule into any desired
position. Since the graphics unit is a two-dimensional screen,
rotation is essential to obtain a good view of the structure.
Furthermore, these routines are useful in locating atoms trapped
in local minima. If such an atom is found, the user can move the
trapped atom to a new position by a MOVE routine found in the
graphics section.
   When the molecule being modelled is in a low strain energy
conformation, the molecular parameters can be listed on an output device or the structure's coordinates can be stored on a
disc file from further processing. In addition a routine has
been interfaced to ADAPT to produce space-filling displays of
structures. The basic algorithm was acquired from a published
report (43) and then interfaced into ADAPT. Figure 2 shows the
type of display that is produced--the upper representation is
with hydrogen atoms suppressed, and the lower one includes the
hydrogens. The only heteroatom in the structure, a hydroxyl

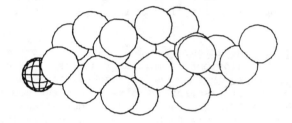

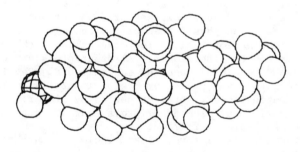

Figure 2.   Space-filling representation of   allopregnan-3α-ol

oxygen, is cross-hatched. The molecule shown is a musk compound.

An automatic version of MOLMEC has also been developed so that data sets with large numbers of molecules can be modelled without continuous supervision. The program consists of an input section, which reads the molecule's connection table and present coordinate matrix from the ADAPT disc files, a minimization section with all output suppressed, and a section which stores the final coordinate matrix. Good models can easily be obtained in this manner. However, before the coordinate matrices can be used for calculating descriptors, the structures are reviewed to make sure that the molecules are in acceptable conformations. Once modelling is complete, geometric descriptors can be derived.

Descriptor Generation. The most important part of SAR studies is the development of molecular structure descriptors. One of the major premises of the approach is that one can find an "adequate" set of descriptors to represent the compounds of interest. The existence of an "adequate" set of descriptors does not necessarily imply that they will be easily found. Thus, descriptor development is the area in which the chemist tests his ingenuity most intensively, bringing to bear on the problem at hand all his insight and knowledge. It is in the area of descriptor development that the most difficult and most potentially rewarding parts of SAR research occur.

There are three general classes of descriptors: topological, geometrical, and physicochemical. Topological descriptors are derived from the topological representation of the structure, the connection table. The geometrical descriptors are derived from the three dimensional model of the molecule. Physicochemical descriptors may be measured experimentally, calculated using a mathematical model, or represented by linearly correlated calculated descriptors. The descriptors that are currently available in ADAPT are as follows:

(a) Fragment descriptors. These include counts of the number of atoms of each type, the number of bonds of each type, the molecular weight, the number of basis rings, and the number of ring atoms.

(b) Substructure descriptors. ADAPT has a substructure searching routine that can be used to develop descriptors. Each of the structures comprising a set of compounds under study is searched for the presence of the substructure of interest. If it is present, then the number of occurrences is computed. If not, then the descriptor is given the value of zero. The substructures to be used are problem dependent and must be found through the application of common sense and experience by the researcher.

(c) Environment descriptors. The information present in the fragment and substructure descriptors indicates the components of the molecular structure. However, the manner of interconnection is missing. Environment descriptors supply information about the connections by coding the immediate surroundings of substructures. To generate an environment descriptor, the molecule being coded is searched for the presence of the substructural fragment that forms the heart of the environment being sought. If no match is found, the descriptor is given the value of zero. If the substructure is found, then the descriptor is computed by performing a path one molecular connectivity calculation on the atoms comprising the substructure, as imbedded within the structure, and in addition the first nearest neighbor atoms. Thus, the value of the path one molecular connectivity represents the immediate surroundings as imbedded within the molecule being coded.

(d) Molecular connectivity descriptors. The molecular connectivity (44) of a molecule is a measure of the branching of the structure. It is formed by summing contributions for each bond in the structure, where the contribution of each bond is determined by the connectivity of the atoms that are joined by that bond. This is the path one molecular connectivity. Higher order molecular connectivities can also be computed by considering all paths of length two, three, etc. These descriptors have been shown in several published reports to be correlated with a number of physicochemical parameters, such as partition coefficients and steric parameters.

(e) Geometric descriptors. Given a three dimensional model of the structures being coded, one can calculate descriptors designed to represent the shape of the molecules. We calculate the three principal moments of inertia and their ratios and the molecular volume.

(f) Electronic descriptors. The first electronic descriptors interfaced into ADAPT were sigma charges calculated by a method due to Del Re (45) and discussed by Hopfinger (46). This approach allows a quick calculation of partial charges on each atom in a molecule. The results were found to be useful in studies of chemical carcinogens (41,42). More recently, we have been interfacing an extended Hückel calculation into our descriptor development routines. We are using the program ICON8 of Hoffmann. This will allow the calculation of a number of reactivity indices previously reported in the literature to be useful in quantum mechanical studies of polycyclic aromatic hydrocarbons (47,48,49). Included will be superdelocalizability, free balance index, bond orders, partial charges for definition of electrophilic or nucleophilic sites, and possibly others.

(g) Partition coefficient. We have developed a routine (50,51) to estimate log P, the logarithm of the partition coefficient between a model lipid phase (usually 1-octanol) and an aqueous phase. It is based on the constructionist approach developed by Leo and Hansch (52). Log P has been shown to be highly correlated with various types of biological activities of organic compounds including pharmaceutical potency, odor quality and intensity, toxicity, pesticidal activity, and bioaccumulation, among others. The log P estimation program provides a true physicochemical parameter for study. Log P values are used for pattern recognition and other analyses. In addition we can calculate correlations between log P and calculated molecular structure descriptors, e.g., molecular volume, molecular connectivity, etc. Thus we can directly test hypotheses regarding the degree to which these calculated molecular structure descriptors are correlated with log P for sets of olfactory stimulants.

In addition to the individual descriptor generation routines, ADAPT has several other supporting routines. There is a general purpose descriptor file management routine that allows the review of any stored descriptor, for example. There is a routine that allows mathematical manipulation of descriptors such as addition, multiplication, logarithmic transformation, exponentiation, autoscaling, etc. There is a routine that allows the user to input descriptors from outside the system so that they can be studied in parallel with the computer-generated descriptors, e.g., gas chromatographic retention indices.

The development of adequate sets of descriptors for the compounds forming a data set comprises the most difficult part of SAR research. With an adequate set of descriptors, the analysis portion of the study is relatively straightforward. With a set of descriptors that is inadequate, one has no choice but to keep searching for better descriptors. Thus, descriptor development for a particular data set can consume quite a lot of time and can be a trial-and-error operation.

Pattern Recognition Analysis. Once each compound in a data set has been represented by a set of molecular structure descriptors, then the analysis phase of the SAR study begins. ADAPT has a variety of pattern recognition and statistical methods available for use. The object of the analysis phase is to find discriminants that separate subsets of the data into the proper categories. That is, one is trying to find mathematical models that will classify compounds as belonging to the active or inactive subset based on the molecular structure descriptors available. This phase of SAR studies is guided by the user in a highly interactive manner in order to search through the available descriptors for the best set.

Pattern recognition is a subfield of artificial intelligence developed largely by electrical engineers and computer scientists. It comprises a set of nonparametric techniques used to study data sets that may not conform to well-characterized probability density functions. A voluminous literature describes the field (e.g., 53,54).

Most of the pattern recognition methods share a set of common properties. The data to be analyzed, here molecular structures of compounds of interest, are represented by points

in a high dimensional space. For a given compound, which is
represented by a given point, the value of each coordinate is
just the numerical value for one of the molecular structure
descriptors comprising the representation. The expectation
is that the points representing compounds of common biological
activity (e.g., compounds with a common odor quality) will
cluster in one limited region of the space, while the points
representing the compounds of another biological activity will
cluster elsewhere. The clusters are regions of high local
density which are relatively far apart from each other. Pattern
recognition consists of a set of methods for investigating data
represented in this manner to assess the degree of clustering
and general structure of the data space.

Parametric methods of pattern recognition attempt to find
classification surfaces or clustering definitions based on
statistical properties of the members of one or both classes
of points. For example, Bayesian classification surfaces are
developed using the mean vectors for the members of the classes
and the covariance matrices for the classes. If the statistical
properties can not be calculated or estimated, then nonparametric methods are used. Nonparametric methods attempt to find
clustering definitions or classification surfaces by using the
data themselves directly, without computing mean vectors,
covariance matrices, etc. Examples of nonparametric methods
would include error-correction feedback linear learning machines
(threshold logic units or perceptrons) and simplex optimization
methods of searching for separating classification surfaces.

Once discriminants have been found that do separate the
data set into the appropriate subsets, then these discriminants
can be used to assess predictive ability. This is usually done
by a round-robin procedure involving leaving out a small number
of data set members to act as "unknown" compounds. When
available, true unknowns can also be input to the system for
prediction of activity.

The final output of the ADAPT-based SAR study is the
identity of the descriptors shown to be correlated with the
biological activity of interest and the discriminants developed.
Study of these can lead to further insights into the biological
activity of interest.

Structure-Activity Studies Using Pattern Recognition
Techniques. A number of studies of the application of pattern
recognition to the problem of searching for correlations between
molecular structure and biological activity have been reported.
A large fraction of the effort in this area must be devoted to
the generation of appropriate descriptors from the molecular
structures available. Areas of study include drug structure-
activity relations, studies of chemical communicants, etc.
Applications of pattern recognition to drug design have been
reviewed by Kirschner and Kowalski (15) and a book has appeared
as well (8).

Studies of Musks

In a previously published study (33) the relationships
between molecular structure and the musk odor quality were
investigated using the computer assisted methods discussed
here. A data set consisting of 60 musk compounds and 240 nonmusk compounds was employed. The 60 musk compounds included 23
macrocyclic, 19 polynitrobenzenes, 11 steroids, 5 gamma-
butyrolactones and two other structural classes. The 240 nonmusk compounds were randomly selected from a larger set of data,
and included were 49 camphoraceous, 44 floral, 32 ethereal, 41
mint, 51 pungent, and 23 putrid compounds. Linear discriminants
were found that could differentiate between the musk compounds
and the nonmusk compounds using 13 molecular structure descriptors. The discriminants were tested on compounds of unknown
olfactory quality and were found to be able to predict odor
quality with very high probability of success.

In the course of that research a common substructural
unit was observed to be present in a large fraction of the
steroid and polynitroaromatic musk compounds. Upon first
inspection, these two classes of musks appeared to have little
in common structurally. However, the substructures shown in
Figure 3 are quite similar. They are not identical because
in the steroid musks the ring portion of the substructure is in
a chair conformation whereas in the polynitroaromatic musks the
ring portion of the substructure is a planar aromatic ring.
However, the degree of similarity demonstrated in the substructure
led us to investigate further the spacial relationships in the
structures of musk odorants.

Geometrical considerations seem to be important for the
presence of the musk quality. To investigate geometrical relationships we have used a set of several hundred musk olfactory
stimulants stored in ADAPT disc files. The data set contains
representative compounds from a number of different structural
classes, e.g., steroidal musks, polynitroaromatic musks, macrocyclic musks, isochroman musks, ortho musks, meta musks, etc.
The six compounds shown in Figure 4 are representative of some
of these structural classes. Each of the musk compounds in
the disc files has had a three-dimensional molecular model
constructed by the MOLMEC routine of ADAPT. In looking closely
at the musk odorants we find relatively invariant spacial relationships between a pair of bonds, one of which contains one
or two heteroatoms (usually an oxygen atom). This relationship
is most easily seen in steroid structure 6 in Figure 4 as
the spacial relationship between the methyl substituent at the
junction of rings A and B and the hydroxy substituent on ring A.
The three-dimensional nature of this spacial arrangement can be
seen in Figure 2. The top side of the steroidal musk, facing
the viewer, is dominated by the two methyl substituents, and
the oxygen atom of the hydroxy group is cross-hatched. This

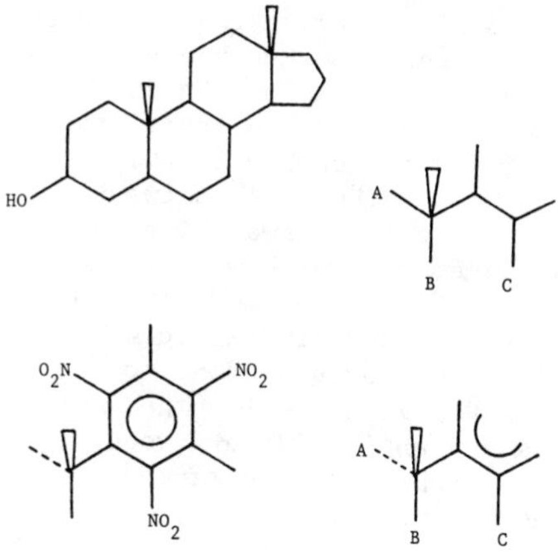

Figure 3. Common substructural unit for steroid and polynitroaromtic musk odorants

1

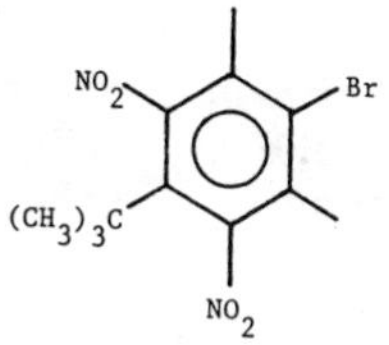

2

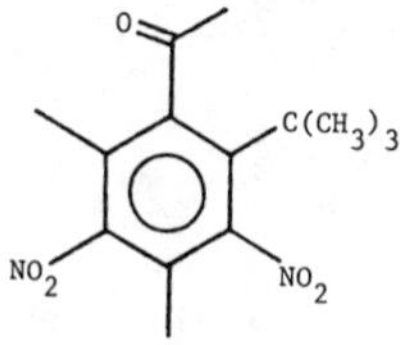

3

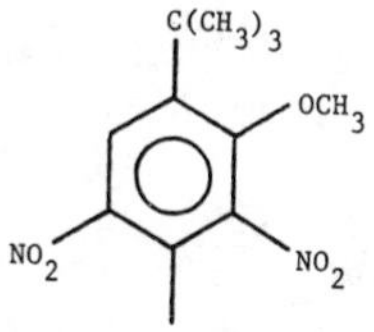

4

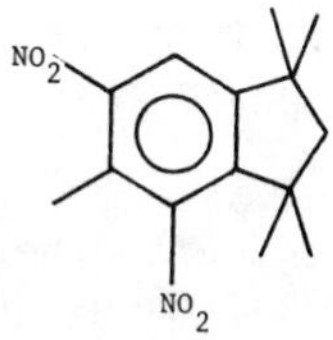

5

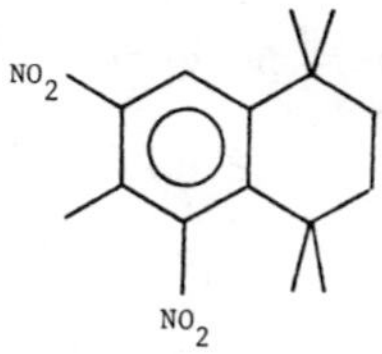

6

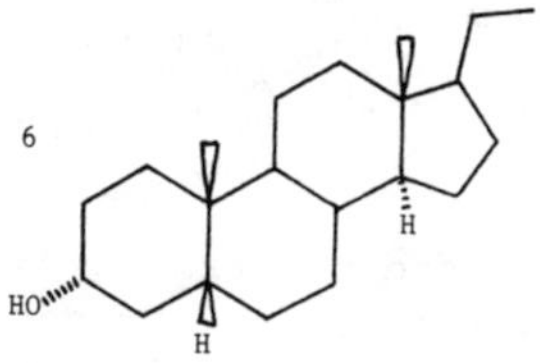

Figure 4. Six musk odorants

spacial relationship can be quantified by measuring the distances between the four atoms involved and the angle formed by the two bonds. For structure 6 of Figure 4 the distance between the methyl substituted and the oxygen atom is 5.51 Å and the distance between the junction of rings A and B and the carbon to which the hydroxy substituent is attached is 3.06 Å. The angle between the two bonds is 141°. We are in the process of searching for the best matches of this spacial arrangement (which we call an "olfactophore" by analogy with the term "phamacophore") of atoms in other known musk compounds. Those compounds in Figure 4 are all musk odorants, and their distances corresponding to those described for structure 6 are given in Table I. The degree of agreement in the distances and angles is very good, suggesting that this portion of these musk compounds may be implicated in the elucidation of the musk odor.

Table I

Geometric Relationships Within the Six Musk Odorants.

Compound	Distance Between Heteroatom and Methyl Group	Distance Between Bases of the Two Bonds	Angle Between the Two Bonds
1	5.71 Å	3.26 Å	153°
2	4.68 Å	2.86 Å	157°
3	4.62 Å	2.64 Å	139°
4	5.51 Å	3.37 Å	140°
5	5.61 Å	3.14 Å	134°
6	5.51 Å	3.06 Å	141°

We are in the process of using the capabilities of the ADAPT system to investigate the properties of the several hundred musk compounds that are stored in ADAPT disc files. Our studies of olfactory stimulants have led us to believe that musk compounds must be relatively large compounds with relatively high lipid solubility. These characteristics are very different from compounds known to be trigeminally active (39) which are relatively small compounds with high aqueous solubility. While this hypothesis was advanced several years ago, we can now estimate the log P values for our musk compounds. The log P values for the six musks shown in Figure 3 are as follows: 6.08, 4.93, 5.27, 5.91, 6.48, and 8.17, respectively. These are certainly compounds that prefer to be in the lipid phase rather than the aqueous phase.

In addition to those studies outlined above, we are now investigating musk olfactory stimulants using another data set. On an ADAPT disc we have a set of 284 musks taken from a chapter on musk compounds in the book by Beets (20), a book by Amoore (55), and a series of papers by Wood (56). The musk chapter by Beets contains 109 compounds that are classified as odorless,

nonmusk, other, or faint and which are of similar structural
types as the musk odorants. We have randomly selected 70 of
these compounds to form a set representative of nonmusk compounds.
In order to keep our data set manageable in size we have taken 140
musks from the three sources above to represent the musk category.
Thus, a well characterized data set of 210 compounds results.
Each of these compounds has been represented by a large number of
calculated molecular structure descriptors, including fragments,
molecular connectivity indices, geometrical descriptors, molecular volume, environment descriptors, log P, and etc. A multiple
linear regression routine has been used to identify descriptors
that are highly correlated with one another. After these identi-
fications were made, then the interrelationships were broken
down by eliminating descriptors in order to produce a set that
does not contain an unacceptable degree of multicollinearity.
After these selections were made, a set of 20 descriptors remained. The log P for each compound was one of these descriptors,
and it was correlated with all other descriptors generated. The
mean correlation coefficient found was 0.282 with the largest
value being 0.83 vs. a path one molecular connectivity descriptor.
The 20 descriptors forming the present set include 3 fragments,
one molecular connectivity descriptor, four molecular connectivity environment descriptors, 9 path environment descriptors, the
molecular volume, log P, and one molar refractivity environment
descriptor. Thus, 14 of the 20 descriptors are substructure
sensitive environment descriptors. In a series of preliminary
pattern recognition studies we have used this set of 20 descriptors to attempt to find linear discriminants that would separate
the 140 musks from the 70 nonmusks. The best results to date
have been obtained with the iterative least squares program,
which developed a linear discriminant that correctly classified
183 out of the 210 compounds for a 87.1% success rate. This
discriminant classified 133/140 or 95.0% of the musks correctly
but only 50/70 or 71.4% of the nonmusks correctly. We are
currently attempting to identify descriptors that will improve
this performance level for classification of the 210 compound
data set. The goal is to generate the most powerful discriminants
possible based on the fewest number of descriptors possible, and
then to use the discriminants to predict new musk compounds.
In performing further studies using these 210 compounds, we
have found sets of descriptors that would support complete
separation between the musk and nonmusk compounds. Full details
of these experiments will be available when the studies are
finalized.

Literature Cited

1. Ariens, E. J., "Drug Design," Vol. 1-8, Academic Press, New York, 1971-78.
2. Burger, A., "Medicinal Chemistry," Part I, Wiley-Interscience, New York, 1970.
3. Bloom, B. and Ullyot, G. E., Eds., "Drug Discovery," American Chemical Society, Washington, DC, 1971.
4. Van Valkenburg, W., Ed., "Biological Correlations -- The Hansch Approach," American Chemical Society, Washington, DC, 1972.
5. Purcell, W. P., Bass, G. E., and Clayton, J. M., "Strategy of Drug Design," Wiley-Interscience, New York, 1973.
6. Martin, Y. C., "Quantitative Drug Design," Marcel Dekker, Inc., New York, 1978.
7. Golender, V. E., and Rozenblit, A. B., "Computer-Assisted Drug Design," Zinathe Press, Riga, U.S.S.R., 1978 (in Russian).
8. Stuper, A. J., Brugger, W. E., and Jurs, P. C., "Computer Assisted Studies of Chemical Structure and Biological Function," Wiley-Interscience, New York, 1979.
9. Dunn, W. J., in "Annual Reports in Medicinal Chemistry," Vol. 8, R. V. Heinzelman, Ed., Academic Press, New York, 1973.
10. Cramer, R. D., in "Annual Reports in Medicinal Chemistry," Vol. 11, R. H. Clarke, Ed., Academic Press, New York, 1976.
11. Hansch, C., in "Advances in Linear Free Energy Relationships," Vol. 2, N. R. Chapman and J. Shorter, Eds., Plenum Press, New York, 1979.
12. Craig, P. N., in "Biological Correlations -- The Hansch Approach," W. Van Valkenburg, Ed., American Chemical Society, Washington, DC, 1972.
13. Richards, W. G., and Black, M. E., in "Progress in Medicinal Chemistry," Vol. 11, G. P. Ellis and G. B. West, Eds., American Elsevier, New York, 1975.
14. Christoffersen, in "Quantum Mechanics of Molecular Conformations," B. Pullman, Ed., Wiley, New York, 1976.
15. Kirschner, G. L. and Kowalski, B. R., in "Drug Design," Vol. VIII, E. J. Ariens, Ed., Academic Press, New York, 1978.
16. Wright, R. H., "The Science of Smell," G. Allen & Unwin Ltd., London, 1964.
17. Wright, R. H., J. Theor. Biol., 1977, 64, 473.
18. Amoore, J. E., "Molecular Basis of Odor," Charles C. Thomas, Springfield, IL, 1970.
19. Beets, M. G. J., in "Structure-Activity Relationships, C. J. Cavallito, Ed., Pergamon Press, 1973.
20. Beets, M. G. J., "Structure-Activity Relationships in Human Chemoreception," Applied Science Publishers Ltd., London, 1978.
21. Theimer, E. T., and Davies, J. T., Jour. Agri. Food Chem., 1967, 15, 6.

22. Theimer, E. T., and McDaniel, M. R., J. Soc. Cosmet. Chem., 1971, 22, 15.
23. Theimer, E. T., Yoshida, T., and Klaiber, E. M., Jour. Agri. Food Chem., 1977, 25, 1168.
24. Teranishi, R., "Odor and Molecular Structure, in Gustation and Olfaction," G. Ohloff and A. F. Thomas, Eds., Academic Press, London, 1971.
25. Laffort, P., Patte, F., and Etcheto, M., Ann. N.Y. Acad. Sci., 1974, 237, 193.
26. Schiffman, S. S., Science, 1974, 185, 112.
27. Schiffman, S. S., Robinson, D. E., and Erickson, R. P., Chem. Senses and Flavor, 1977, 2, 375.
28. Dravneiks, A., Ann. N.Y. Acad. Sci., 1974, 237, 144.
29. Dravnieks, A., A.C.S. Symp. Ser., 1977, 51, 11.
30. Boelens, H., "Molecular Structure and Olfactive Properties, in Structure-Activity Relationships in Chemoreception," G. Benz, Ed., Information Retrieval Ltd., London, 1976.
31. Greenberg, M. J., J. Agri. Food Chem., 1979, 27, 347.
32. McGill, J. R., and Kowalski, B. R., Anal. Chem., 1977, 49, 596.
33. Brugger, W. E., and Jurs, P. C., J. Agri. Food Chem., 1977, 25, 1158.
34. Stuper, A. J., and Jurs, P. C., Jour. Chem. Infor. Comp. Sci., 1976, 16, 99.
35. Brugger, W. E., Stuper, A. J., and Jurs, P. C., Jour. Chem. Infor. Comp. Sci., 1976, 16, 105.
36. Stuper, A. J., Brugger, W. E., and Jurs, P. C., A.C.S. Symp. Ser., 1977, 52, 165.
37. Stuper, A. J., and Jurs, P. C., Jour. Amer. Chem. Soc., 1975, 97, 182.
38. Stuper, A. J., and Jurs, P. C., Jour. Pharm. Sci., 1978, 67, 745.
39. Doty, R. L., Brugger, W. E., Jurs, P. C., Orndorff, M. A., Snyder, P. J., and Lowry, L. D., Physiol. and Behavior, 1978, 20, 175.
40. Jurs, P. C., Chou, J. T., and Yuan, M., Jour. Med. Chem., 1979, 22, 476.
41. Cour, J. T., and Jurs, P. C., Jour. Med. Chem., 1979, 22, 792.
42. Yuan, M., and Jurs, P. C., Toxicology and Appl. Pharmacology, 1980, 52, 294.
43. Smith, G. M., and Gund, P., Jour. Chem. Inf. Comput. Sci., 1978, 18, 207.
44. Kier, L. B., and Hall, L. H., "Molecular Connectivity in Chemistry and Drug Research," Academic Press, New York, 1976.
45. Del Re, G., Jour. Chem. Soc., 1958, 4031.
46. Hopfinger, A. J., "Conformational Properties of Macromolecules," Academic Press, New York, 1973.
47. Berger, G. D., Smith, I. A., Seybold, P. G., and Serve, M. P., Tetrahedron Letters, 1978, 231.

48. Smith, I. A., Berger, G. D., Seybold, P. G., and Serve, M. P., Cancer Research, 1978, 38, 2968.
49. Smith, I. A., and Seybold, P. G., Int. Jour. Quan. Chem.: Quan. Biol. Symp., 1978, 5, 311.
50. Chou, J. T., and Jurs, P. C., Jour. Chem. Infor. Comp. Sci., 1979, 19, 172.
51. Chou, J. T., and Jurs, P. C., in "Physical Chemical Properties of Drugs," S. H. Yalkowsky and A. Sinkula, Eds., Marcel Dekker, Inc., New York, in press.
52. Hansch, C. and Leo, A., "Substituent Constants for Correlation Analysis in Chemistry and Biology," Wiley-Interscience, New York, 1979.
53. Nilsson, N. J., "Learning Machines," McGraw-Hill, New York, 1965.
54. Tou, J. T., and Gonzalez, R. C., "Pattern Recognition Principles," Addison-Wesley, Reading, MA, 1974.
55. Amoore, J. E., "Molecular Basis of Odor," Charles C. Thomas, Springfield, IL, 1970.
56. Wood, T. F., "Chemistry of the Aromatic Musks," The Givaudanian, 9 papers, 1968-1970.

RECEIVED October 27, 1980.

Structure Recognition as a Peripheral Process in Odor Quality Coding

ALFRED A. SCHLEPPNIK

Monsanto Company, Biomed Department, 1800 North Lindbergh Blvd., St. Louis, MO 63166

The odor quality of a compound is, according to BEETS (1), defined intrinsically by the chemical structure: Odorant molecules encode the structural modalities of the stimulant molecule in a transduction process, which, taking all changes of orientation and conformation into account, produces informational modalities. The latter are expressed as topologically defined structural features of high variability and complexity.

Odor/Structure Correlation attempts to elucidate the mechanisms which mediate the information transfer from structural features of a molecule to a corresponding information pattern. The latter originates in olfactory neurons and is encoded in nerve impulses. It is projected for further analysis, discrimination and recognition to the higher olfactory centers of the CNS. This information transfer includes the transduction process which converts chemical to electrical signals.

Many odor theories have been proposed in the past, attempting to explain the multitude of often very complex phenomena observed in human olfaction. Most of them were only partially, if at all, successful. Nevertheless, slowly a consensus developed and today it is generally assumed that the primary process of chemoreception takes place at the cell membrane of a sensory neuron and involves physical contact of the stimulant with potential or actual receptor sites which could be either specialists - reacting only with one structural class - or generalists which would react with a multitude of structural classes.

0097-6156/81/0148-0161$05.00/0

Then, interaction of the stimulant molecule with the receptor site
regardless of the nature of the processes involved, has to achieve
the following results:

    a)   Graded transduction of a chemical into an electrical
        signal (Intensity Grading)

    b)   Transcription of all or significant parts of the struc-
        tural modality of the stimulant molecule into a set of
        informational modalities which are combined in a precise
        and specific "Odor Information Pattern" (Quality Coding)

    c)   Amplification of the primary energy gained by adsorption
        of a few stimulant molecules to a level high enough to
        trigger the electrogenic processes involved in signal
        generation (Depolarization of the sensory neuron, firing
        of a spike)

    d)   High speed of the total process to create a potential in
        a few 100 msec.

    e)   Do all this without involving the stimulus in any chemi-
        cal changes, but release it unchanged rapidly after ter-
        mination of the transduction process.

In this communication the focus is on b): Odor/Structure Correlation.

Most, perhaps all of the odor theories advanced so far made the
assumption that the transcription of structural information encoded in the stimulant molecule into an odor information pattern
is an integral process: One odorivector (AMOORE, 2) interacts with
one receptor site and this interaction results in transcription of
all structural components simultaneously into their corresponding
informational modalities. However, observation tells us that olfactory information is inherently complex: Ambergris for instance
is described (OHLOFF, 3) by six distinctly different notes. This
would imply that in an integral process of the peripheral molecular interaction one single neuron has to detect at least six different profiles with six different receptor sites and project the
informational modalities intact to the higher centers.

Since the single bit of olfactory information is one spike of the
olfactory neuron which is independent of the number and qualities
of the detector sites an insurmountable problem for quality coding
arises. One way to avoid this problem is simply to deny the existence of specific receptor sites and specialized detector cells in
AMOORE's terms and replace the specialised concept with a "General
Concept" in which quality coding is achieved through a spatial
distribution of collections of a large number of structurally
different generalist receptor sites which would interact with the
stimulant molecule in all its orientations and conformations. In
this way all structural features of the stimulant molecule - the
structural modalities - would be converted into informational modalities distributed over an information pattern with more or less
distinct topological characteristics. Therefore the profile is not

expressed at the molecular or microscopic, but at the macroscopic level, as defined areas of the olfactory epithelium. However it has to be noted that again, even in this diffuse pluriform interaction scheme the integral process is used: One stimulant molecule interacting with one generalist receptor site is sufficient for quality coding.

The alternative to the integral process is a differential process of the type of a "Multiple Profile - Multiple Receptor Site" interaction first suggested by POLAK Jr. (4). In a system of this kind the profile and the receptor site have to be sterically complementary like a substrate to an active site of an enzyme; or a drug molecule to its specific receptor site; or a hormone to its complementary regulatory site of a membrane bound adenyl cyclase system. It is characteristic that almost all life processes are regulated by interaction of chemical messenger molecules with specific sites of a tertiary protein structure. Staying within the well established and accepted principles of molecular biochemistry and assuming that there is indeed no drastic difference between the peripheral processes of substrate/enzyme-, drug/specific receptor site- and stimulant/specific receptor site interactions one can postulate that the tertiary protein structure - the receptor site - is part of the regulatory subunit of an adenyl cyclase system. The same regulatory subunit could contain a second regulatory site.

Adenyl cyclases are highly complex enzyme systems consisting of several interacting subunits. The system described above contains a subunit with two regulatory sites: One for the odorivector which acts as an activator for the catalytic site of the adenyl cyclase system imbedded in a second subunit. The other regulatory site in the first subunit then can act as an allosteric regulatory site for activators or inhibitors and in this manner regulate the conformation of the specific odorivector receptor site.

The second subunit of the adenyl cyclase system is the catalytic subunit. It forms a stable binary complex with the magnesium salt of adenosine triphosphate (ATP) in its resting state. The stability of the binary complex is caused by the complexed ATP-molecule which acts as an undersized blocking agent. Arrival of an odorivector at the activating allosteric site in the regulatory subunit and the subsequent complex formation of the odorivector with the "Detector subunit" results in conformational changes of the detector subunit which are communicated through cooperative effects to the catalytic subunit. The latter then can adapt again through conformational changes of the tertiary structure the catalytic site to the substrate -ATP- which it already contains. This "Induced Fit" (KOSHLAND, 5) activates the catalytic site, the catalytic (enzymatic) reaction takes place very rapidly and ATP is converted to 3',5'-adenosine monophosphate. This "cyclic adenosine

monophosphate (cAMP)" is the second messenger: still a chemical
signal, the ubiquitous information carrier in regulatory enzymatic processes.

Adenyl cyclase systems isolated from mammals contain as a third
component an additional guanosine triphosphate specific subunit
and in all probability even more components whose function and
structure are not known as yet. It is interesting to note that
adenyl cyclases, Na-K-activated adenosine phosphatases (ATPases)
have been located in the membrane of olfactory neurons; and cAMP
was found to have the highest concentration in man in the olfactory mucosa.

The second messenger, cAMP, couples the adenyl cyclase which functions as a "Detector Enzyme" to another membrane bound enzyme, a
Na-K-ATPase which operates as an ion pump which moves ions in ac-
active transport against their concentration gradient. Changes of
the activity of the ATPase produce changes of the membrane potential. Therefore regulation of the ion pump by the second messenger
-cAMP- produces regulation of the membrane potential. Furthermore,
assuming that both the detector enzyme (adenyl cyclase) and the
transducer protein (ATPase) are monomers of a heterogenic polymer-
ic enzyme system arranged in a two dimensional pattern in which activation of one coupled enzyme pair would, by positive cooperative
effects, activate a large number of acceptor units (transducer +
detector enzymes) not only a single Na/K pump (the transducer enzyme), but a very large number of Na/K-pumps would be regulated.
As a consequence of such a mechanism a powerful amplification factor would be introduced: The two dimensional multienzyme system
operates like a bioamplifier.

Arrival of a single odorivector molecule at its complementary specific receptor site consequently leads to partial depolarization
of all transducer cells involved in the bioamplifier. The resulting change in membrane potential has been observed as the "Generator Potential". If it builds up high enough it triggers a third
enzyme system which instantly depolarizes the olfactory neuron.
The resulting change in membrane potential is a single nerve im-
pulse, a spike. Since this third enzyme system produces a strong
signal on reception of a weaker one it works as a true transponder
which indicates by generation of a spike that a generator potential had reached a critical level. The spike is the single bit of
chemoreceptory informational modality transcribed from structural
modalties of the odorivector.

In short, ligand formation of one odorivector molecule with a receptor site having a complementary structure to structural elements of the odorivector would result in formation of a single bit
of chemoreceptory information. The acceptor system is a modular
system in which the transducer and the transponder can remain un-

changed and only a change of the detector enzyme in the regulatory subunit of the detector enzyme is required (and sufficient) to
provide for the accommodation of a practically unlimited number
and variety of structural features of the odorivector through complementary structural features of the receptor site.

At first glance this seems to resurrect the old "Specialised Concept" in which a specific receptor site for a "typical odorivector structure" and its congeners would engage in ligand formation
in an integral process. This would be in sharp contrast to experimental results obtained in single cell electrophysiological studies. These demonstrate that at least in vertebrates the olfactory
neurons are not specialists, but GENERALISTS AS FAR AS THE OVER-
ALL STRUCTURE OF THE MOLECULE IS CONCERNED: They interact with
a multitude of structurally different odorivectors. This observation insinuates that not the total sum of all structural features
of the odorivectors, NOT THE OVERALL STRUCTURE, is encoded, but
a SPECIFIC PARTIAL STRUCTURAL FEATURE which may very well be part
of many otherwise totally different overall structures of odorivectors.

M. G. J. BEETS (6) has introduced the term "Profile" for this type
of partial - or submolecular - structure. This principle and the
term were adopted, but in the system discussed now - the ENZYME
MODEL OF OLFACTION - the meaning of "profile" was defined more
sharply. In it the term "profile" describes a limited number of
well defined substructures of the odorivector. In the EMO a profile of the odorivector consists of a three dimensional spatial
arrangement of a sequence of atoms in a well defined overall geometry. It can be present explicitly, preformed if the odorivector or a significant part of it has a rigid structure with practically almost no conformational freedom. However the profile can
be contained implicitly in odorivector molecules with varying degrees of conformational freedom. Such molecules have either "eloquent" structures with high degrees of conformational freedom,
capable of expressing their structural modalities in many different ways; or flexible molecules with a limited range of conformational freedom in which one or a few conformations would be vastly preferred and others excluded. It follows that eloquent, and
to a lesser degree, flexible molecules can, all other sterical
requirements provided, assume the same profile as one preformed
in a rigid odorivector structure. However, with increasing conformational freedom the probability of assuming a specific profile diminishes rapidly.

It furthermore follows that a profile may constitute only a significant part of the overall structure of the odorivector molecule: either a shape - the Van der Waals molecular outline proposed by AMOORE - which can degenerate to a molecularly defined
plane; or it can be a functional group in the traditional sense

of "Osmophores" (RUZICKA's odor theory, 7) which may contribute
stereoelectronic features, such as Pi-electron clouds etc.

In discussions of Odor/Structure correlations the odorivector
can be treated as a collection of a number of explicit or implicit profiles which are either directly connected or imbedded in
a larger "frame structure". The resulting molecular weight of
the structures generated in this way has little influence on the
interactions with olfactory receptor sites as long as the fuga-
city of the odorivector is high enough to allow a sufficient number of odorivector molecules to reach receptor sites. Small molecules with a molecular weight of less than 100 Daltons display
in addition to their normal interaction with complementary receptor sites projecting signals into the olfactory nerve, strong interactions with a branch of the trigeminus nerve, causing the
well known effects of irritation and interference with odor perception (CAIN and MURPHY, 8).

Typical odorivectors - most "odorant molecules" - have a molecular weight in the range of 100 to about 350 Daltons. They contain therefore enough "skeletal" atoms to build frame structures
for explicit functional groups or distinct shapes. It is this
type of odorivector with one functional polar group attached to
or imbedded in an often very complex frame which is the one most
commonly encountered. Since the frame part can potentially contain a plurality of profiles the total odorivector itself can
carry a vast amount of structural information. Conversely, the
small molecules with a molecular weight below 100 Daltons have
only very small "frames", if any at all, and consequently carry
only a limited amount of structural information beyond their inherent trigeminus irritant contribution to the overall sensory
perception.

In any case, whatever amount of structural modality may be contained in the odorivector structure has to be transcribed totally
or partially in the transduction process. More precisely, this
transcription process has to be effected in the peripheral interaction of the odorivector with the receptor site leading to ligand formation. The resulting complex is bound by weak and reversible bonds, such as hydrogen bonds or Van der Waals forces.
In most cases the receptor site is the proton donor, most likely
through free thiol groups. In some special cases the reverse
process, in which the odorivector acts as a proton donor, may be
operational.

In order to achieve weak bond formation the ligand has to fit into the receptor site in such a way as to bring weak bond forming
sites of the odorivector and the receptor site within striking
distance. This is the same process as the one assumed in drug/
receptor interactions. It was recognized in molecular pharmacolo-

gy that the ligand could be construed to consist of an "Affinity Part" and an "Intrinsic Activity Part" (ARIENS, 9). This concept is loosely comparable to the description of a "normal odorivector molecule" as a "frame plus one polar functional group".

The affinity part determines the ease with which the complementarity of profile and receptor site is achieved. This is dependent of the equlibrium structure of both the profile and the receptor site and the energy required to change the conformation of either component or both.

The "Intrinsic Activity Part" determines the ease of weak bond formation of the functional group with the active group of the receptor site. It has been suggested that this weak bond formation occurs as the first step and provides a pivot for the affinity part which encodes the "Shape profiles" and therefore has to assume the proper orientation before the complex formation is finished by induced fit of the receptor site. It is noteworthy that in such cases weak bond formation to the functional group does not encode the structural modality of the functional group, but of that of the shape. Functional groups have their own, in most cases sterically less demanding, specific receptor sites.

From this follows that increased size and sterical complexity of the frame (affinity moiety) potentially provides a larger number of shape-profiles. Assuming that all these sterical modalities are expressed in informational modalities the contribution of functional groups becomes proportionally less distinctive, and, given a sufficiently effective steric hindrance of the functional group may render its informational modality in the overall odor information pattern negligeable. In this way the older "Functional Group Odor Theory" of RUZICKA (1920) and the "Stereochemical Theory of Odor" of AMOORE (1962) are reconciled: Both are totally compatible with the "Enzyme Model of Olfaction" and deducible from the general molecular requirements of ligand formation.

Odorivector molecules can contain an almost unlimited number of profiles. Of these are only a few explicit, but with increasing conformational freedom a rapidly increasing number of implicit ones are potentially possible. This raises the question about the number of complementary receptor sites necessary to deal unambiguously and efficiently with the transcription of structural into informational modalities. The concept of multiple profile - multiple receptor sites provides no indication how the actual number of receptor site types could be deduced. However the minimum number required to encode the total olfactory spectrum perceived by man can be estimated by means of basic principles of information theory. For that a few simple assumptions have to be made:

1) Each receptor site produces only one informational moda-
    lity, a monoosmatic component.
2) Each olfactory neuron contains exclusively or vastly pre-
    dominantly receptor sites specific for only one profile.
3) The contributions - nerve impulses - of the individual
    active neurons are summated in the next higher center,
    the glomeruli of the olfactory bulb. If the combined ac-
    tivity of 20.000 - 25.000 olfactory neurons, which all
    feed into one glomerulus, excede a threshold value, the
    glomerulus is activated - turned on to produce a signal
    which indicates a specific monoosmatic component.
4) All specific monoosmatic components are combined in still
    higher centers to produce an "Odor Information Pattern".
5) Each discernible odor has a specific unique individual
    odor information pattern.

It has been observed that the discriminatory capabilities of human olfaction are tremendous: It was estimated that an untrained
person could differentiate up to ten million odors, perhaps even
significantly more than that. Information theory then shows that
in order to encode the qualities of ten million odors in a simple
binary mode (Monoosmatic components on or off, their intensity,
albeit important, is in this connection disregarded) only 24 to
27 specific profiles, disregarding possible and probable redun-
dancies, and therefore the same number of complementary receptor
sites would be required. Assuming furthermore that said redundancy, in which the informational modalities of two different specific receptor sites of two different olfactory neurons are conflu-
ent in one collector cell and therefore contribute to the expression of only one monoosmatic component is indeed operational it
becomes necessary to increase the total number of types of specific receptor sites to 24-30. This means that only 24-30 specific
detector proteins are required for structure recognition in the
transduction process. This compares to about 4000 enzyme systems
in different stages of activity estimated to be present in a cell
any time.

The next question arising is that about the minimum number of monoosmatic components required to encode an odor quality. It has
been recognized by BEETS that an inherent "Principle of informational complexity" makes the perception of even a single odorant
molecular species informationally complex, even if the odor information pattern is dominated by the terminal derivative (monoosmatic component) of a single chemoreceptory modality. But there has
to be something like a minimum complexity still. In terms of the
EMO there must be a minimum number of monoosmatic components essential to produce a minimal odor information pattern. Again,
since this problem is not in the domain of peripheral processes,
the Enzyme Model of Olfaction cannot provide an answer. However,
experimental results obtained by POLAK (10) indicate that one

single informational modality does not encode a quality but only
signals the presence of an odorant.

As a consequence the minimum number of monoosmatic components required to encode an odor quality is two. This seems to rule out
the concept of primary odors. However, taking into account the
relative intensities of the monoosmatic components one could expect that an odor profile with two monoosmatic components of which
one dominates decisively would signal an odor quality approaching
the simplicity of a primary odor.

Arrival of the odorivector in its prefered orientations and conformations at the olfactory epithelium leads to simultaneous complex formation of many odorivector molecules through different
profiles contained as structural modalities in the overall structure with their corresponding complementary receptor sites. This
leads to signal generation and signal modification and produces
an odor information pattern in which each monoosmatic component
indicates the presence of a distinct chemical structural feature.
Consequently the odor information pattern denotes not only a well
defined odor quality, but, by signaling the presence of specific
functional groups, characteristic shapes and electron distribution, expresses an abridged qualitative analysis of the odorivector.

Therefore in any attempt of odor-structure correlation not the
total (or overall) structure of the molecule should be considered
but the individual contributions of the molecular profiles. Perhaps this could be done by a combination of computer assisted conformational analysis of the odorivectors which would provide information about the nature of the explicit and implicit profiles
as well as the probability of the formation of the latter, with
multidimensional scaling of the highly processed information the
odorivectors deliver.

Furthermore the odorivectors could be treated the same way, with
the same methods, as drug molecules are in QSAR (Quantitative
Structure Activity Correlation). A computerized approach to biochemical quantitative structure-activity-correlations was introduced by the HANSCH APPROACH (11). Definition of all the essential profiles, those capable of being expressed in monoosmatic components, would afford the foundation on which an algorithm for
the calculation of odor quality based on the chemical structure
of the odorivector conceivably could be designed.

Up to this point only speculations have been presented. They were
based on the assumption that the peripheral process in olfaction
is mediated by specific receptor sites of a group of membrane
bound adenyl cyclases and that the Multiple Profile - Multiple Receptor Site concept is viable. If these assumptions **are correct**
the following extrapolations could be made:

1) Adenyl cyclases are regulatory enzymes and themselves
   subject to allosteric regulation of their specific receptor site. Therefore it should be possible to regulate the activity, and hence the sensitivity of the detector subunit.

2) Regulation of the detector system of a specific regulatory subunit would result only in the change of the contribution of one monoosmatic component - the one whose
   detector sensitivity is changed. Therefore the odor information pattern would remain unchanged except for the
   contribution of that single monoosmatic component whose
   corresponding receptor site has been regulated (activated or inhibited).

3) Regulation of a single monoosmatic component could lead
   to noticeable changes in odor quality.
   Inhibition would reduce or even eliminate the contribution of a dominant or significantly modifying monoosmatic component and thereby cause a noticeable antagonistic effect. Further reduction of a minor monoosmatic
   component or its elimination would go in all probability undetected.
   Activation could raise the contribution of a minor monoosmatic component to either modifying or dominant
   status and thus create a noticeable synergistic effect.
   Both antagonistic and synergistic effects are very common in multicomponent odorivector systems and are well
   known to experienced perfumers.

4) These observed synergistic and antagonistic effects indicate that the regulatory activity has to be encoded
   in an odorivector present in the mixture, in all probability in the same way as the activators of the detector adenyl cyclase - as an "Active Profile"

In terms of established principles of enzyme chemistry there is
no difference between the interaction of a molecular profile
with its complementary receptor site and that of an active profile with its corresponding complementary regulatory site. In
both cases normal ligand formation through weak bonds takes
place.

The active profile can be a specific regulatory one which does
not interact with a normal detector site. Consequently it would
not produce a monoosmatic component and the regulatory activity
would be independent of the intrinsic odor of the regulatory molecule itself. The other possibility of course is that the same

molecular profile could direct ligand formation with a complementary receptor site and contribute in this way to the formation of the corresponding monoosmatic component. Beyond that it could direct ligand formation with a regulatory site and in this way interfere with the transcription of the structural information modalities of a copresent odorivector. Finally, an active profile in an odorivector could interfere with its own transcription.

In all of these examples of regulation of transcription of profiles by allosteric regulation of receptor sites by active profiles THE DUAL NATURE OF ODORIVECTORS manifests itself. This is a new principle postulated to be pertinent in all mixtures of odorivectors. In its most extended scope this principle states that all odorivectors have two functions: To display their own intrinsic odor and at the same time act as regulator in the odor perception of a copresent odorivector. The latter is achieved by allosteric regulation in a peripheral process.

The Dual Nature of Odorivectors explains the observed nonlinear additivity of odors. In odor mixtures the contribution of each component is not necessarily the odor quality it would display if presented as a single odorant - the intrinsic odor - but an odor quality which is changed by the allosteric regulation caused by a co-present odorivector. The extent of this change is a function of the concentration of the regulatory odorivectors present.

The concept of the Dual Nature of Odorivectors furthermore explains all observed irregularities, synergistic and antagonistic effects at least in part by assuming the causative processes take place at the periphery and not exclusively at the CNS-level as has been generally assumed so far.

That this peripheral interaction of odorivectors is a reality and not just a postulate resulting from lengthy speculations has been confirmed by statistically significant experimental proof obtained in malodor/"antimalodor"-interaction studies (12), and on a more general base, in odor/odor-interactions. These results give implicit proof that specific receptor sites for molecular and active profiles exist.

The "Antimalodors" (AMALs) mentioned above were discovered in a chance observation in 1968 (13). In a routine screening program of new aroma chemicals it was found that several new compounds had the unique property to suppress the perception of malodors caused by molecules which have pronounced proton donor or proton acceptor properties. The most commonly encountered malodors belong in this group: lower fatty acids, phenols, mercaptans, amines etc. Even more important was the observation that these "Antimalodors" produced a very specific counteraction effect which did not interfere with the perception of all other odor qualities. On top of that the AMAL-activity was highly specific and rapidly and totally reversible. In other words, the AMALs induced specific reversible anosmia. Later on, in an extensive screening program it was demonstrated that the AMAL-activity of odorivectors was totally independent of their intrinsic odor qualities. Psychophysical precision measurements finally showed that at the very high dilutions of their application levels the AMALs were all subthreshold.

All these observations combined make it obvious that we had indeed a regulatory interaction and not one of the traditional malodor counteractions such as simple overpowering or masking. This view is supported by the fact that the antimalodors have no structural similarity to most of the common malodors but present "normal" aroma chemical types, with molecular weights well above 100 Daltons and different degrees of polarity.

In contrast the malodors - proton donors or proton acceptors with no exception - are all small molecules with molecular weights well below 100 Daltons, they are highly polar compounds and have little or no steric requirements. They share no structural features (Methyl mercaptan - trimethyl amine - isobutyric acid - phenol) and all are strong irritants.

These observations lead to two very important conclusions:

a.   The oberved inhibition of malodor perception cannot be caused by competitive inhibition. In such a mechanism the AMAL-molecule would block the common receptor site for all prototropic malodors and the steric requirements for the malodors and the AMALs would have to be very similar in order to make ligand formation of both types with the same receptor site possible. As has been pointed out already exactly the opposite is the case: AMALs with their sterically well defined active profiles require sterically equally well defined receptor sites for ligand formation, whereas the receptor site for the common entity of all malodors has no sterical requirements at all.

This points to allosteric regulation of the critical receptor site common to all malodors. The allosteric site undergoes ligand formation with the AMAL active profile. This process causes changes in the conformation of the regulatory subunit of the detector enzyme which alters the overall geometry of the sterically indifferent critical receptor site to such an extent that its activity for ligand formation with malodors is decreased or totally inhibited. Consequent-

ly, if the AMAL reaches its regulatory receptor site
prior to the arrival of a malodor molecule at the
critical receptor site the ensuing conformational
change of the latter makes ligand formation difficult
or impossible, the transduction process is slowed
down or totally inhibited. Consequently the contribution of the monoosmatic component common to all
malodors is reduced or totally eliminated. As a result the malodor cannot be perceived in its original
intensity or not at all.

b. The receptor site common to all prototropic malodors
- the "critical" receptor site - then has to have the
ability to recognise the presence or absence of a
proton donating group (-COO$\underline{H}$, Ar-O$\underline{H}$, R-S$\underline{H}$, electro-
philes) or a proton accepting group (RR'$\overline{\text{R}}$"N, nucleo-
phile). Several structures which would have this pro-
perty and can be assembled from functional groups common in proteins, such as carboxyl- , mercapto- or primary amino groups. A "Reinforced ionic bond" formed
from a carboxyl- and a primary amino group would give
through formation of two hydrogen bonds between two
hydrogens on nitrogen and the two oxygens in the carboxylate anion a resonance stabilised six membered ring
system. A proton donor would donate a proton to the
ring system which would then open to give an ammonium
carboxylic acid; whereas a proton acceptor would ac-
cept a proton and break the resonance stabilised six
membered ring to give an amino carboxylate anion.
Since both functional groups can be part of distant
amino acids brought into proximity in the tertiary
protein structure formation of a resonance stabilised
reinforced ionic bond could stabilize one conformation and its ring fission could bring about profound
conformational changes.

The final unequivocal experimental proof that the observed effects were indeed peripheral ones was obtained in psychophysical
experiments. Tertiary butyl mercaptan was used as the target in
a monorhinal presentation. Its perceived odor intensity remained
unchanged when in a dichorhinal experiment the contralateral na-
ris was exposed to a very low intensity of 4-cyclohexyl-4-methyl-
2-pentanone (CMP, $\underline{13}$). However, in agreement with established
crossover additivity, the total perceived overall odor intensity
showed a small, but statistically significant increase. Then the
two separate odorant streams of the dichorhinal experiment were
combined and the mixture of malodor (t.-butyl mercaptan) and
AMAL (CMP) presented to the subjects again. Perceived overall
intensity was reduced by 74% and perceived malodor intensity by
85% at a significance level of 5%.

Similar experiments with other malodor/AMAL-combinations gave the same results. For example when isovaleric acid and CMP were used perceived overall intensity in the monorhinal presentation of the mixture was reduced by 64% in comparison with the dichorhinal presentation; and perceived malodor intensity by 96%.

When linalyl acetate was used as the target in place of the prototropic malodors in the same experimental protocol no difference between monorhinal presentation of the target and mono- and dichorhinal presentation of target and AMAL (CMP) was observed.

These results cannot be explained with any of the older theories of olfaction whereas the Enzyme Model of Olfaction not only can do that effortlessly, but actually allows to predict these effects on the basis of generally accepted principles of molecular biochemistry. The concept of "STRUCTURE RECOGNITION AS PERIPHERAL PROCESS IN ODOR QUALITY CODING" represents only the special application of a more general mechanism of structure recognition in peripheral processes to the problems of quality coding in olfaction.

Acknowledgement
The author gratefully acknowledges the support by colleagues and management of MONSANTO FLAVOR/ESSENCE from 1968 - 1978. The research initiated at M F/E is now being continuated by Bush Boake Allen, Inc, Montvale, NJ.

Abstract

Interaction of odorivectors with receptors leading to signal generation and subsequent formation of an odor information pattern composed of a limited number of monoosmatic components can be visualized to proceed by either an integral or differential process. The integral process is molecular: The total odorivector molecule is involved in a single interaction which triggers a transduction process capable of producing a multicomponent information pattern. The differential process is based on a multiple profile/multiple receptor site mechanism: Many odorivector molecules interact independently through different submolecular profiles with complementary specific receptors in profile specific transduction processes, each of which leads to formation of a specific monoosmatic component of the final odor information pattern. In this mechanism therefore specific regulation of formation of monoosmatic components should be possible and should lead to distinct changes in perceived odor quality caused by the resulting selective synergistic and antagonistic effects. Implications of the concept of the differential process and experimental results of selective specific antagonistic effects are discussed in this communication.

Literature Cited

1. BEETS, M. G. J. "Structure-Activity Relationships in Human Chemoreception"; Applied Sciences Publishers Ltd.: Barking, Essex, England, 1978.
2. AMOORE, J. E. "Molecular Basis of Odor": American Lecture Series, Publication No. 773; Ch. C. Thomas, Springfield, Illinois; 1970.
3. OHLOFF, G. "Relationship between Odor Sensation and Stereochemistry of Decalin Ring Compounds" in G. OHLOFF and A. F. THOMAS, Eds. "Gustation and Olfaction"; Academic Press New York-London; 1971; p. 178
4. POLAK, E. H., J.Theor.Biol. 1973, 40, 469
5. KOSHLAND Jr., D. E., Ann.Rev.Biochemistry, 1968, 37, 359
6. BEETS, M. G. J., "Molecular Approach to Olfaction" in: E. J. ARIENS, Ed., "Molecular Pharmacology"; Academic Press New York-London; 1964, Vol. II, p. 1
7. RUZICKA, L., Chemikerztg. 1920, 44:19, 129
8. CAIN, W. S.; MURPHY, C. L., Nature, 1980, 284, 255
9. ARIENS, E. J., Arch.int.pharmacodyn. 1954, 99:1, 32
10. POLAK, E. H., J.Am.Oil Chem.Soc., 1968; 45, 680A
11. GOULD, Robert F., Ed. "Biological Correlations - the Hansch Approach", Advances in Chemistry Series 114; American Chemical Society, Washington, D.C.; 1972
12. SCHLEPPNIK, Alfred A., to be published
13. SCHLEPPNIK, Alfred A., VANATA, S. G., U.S. 4,009,253, Feb. 22, 1977

RECEIVED October 29, 1980.

The Dependence of Odor Intensity on the Hydrophobic Properties of Molecules

MICHAEL J. GREENBERG

The Quaker Oats Company, 617 West Main Street, Barrington, IL 60010

The quantitative approach to understanding biological activity depends upon being able to express structure by numerical values and then relating these values to corresponding changes in activity.

Relatively little work in this area has been reported on how odor intensity is dependent upon odorant physical chemical properties. Davies and Taylor (1) related threshold to the cross-sectional areas and adsorption constants at an oil-water interface of the odorant molecules. However, these observed and calculated thresholds frequently varied by +1 logarithmic units and sometimes as much as +2.5 units. Guadagni et. al. (2) related molecular weight with the odor threshold values of aliphatic aldehydes in water. Beck (3) assumed that a factor determining an odorant's threshold is its volume, shape, and axis (produced by the odorant's functional group "anchored" at a receptor site) around which the molecule rotates. In another study odor thresholds were related to odorant air water partition coefficients, hydrogen bonding, molecular volume and polarizability by Laffort (4). Laffort et. al. (5) also correlated odor intensity with GLC retention parameters.

In still another study Dravnieks (6) correlated 14 structural features with odor threshold and suprathreshold data. More recently Dravnieks (7) correlated odor intensity equivalent to 87 ppm (Vol/vol) of 1-butanol with 20 structural features represented by Wiswesser line notation. The molecular weight term, $(\log mw)^2$, was reported to be the most statistically significant term.

The use of computer techniques in the correlation of biological activity with substrate physical-chemical properties has received much attention in the area of medicinal chemistry. The use of these techniques, denoted Quantitative Structure Activity Relationships (QSAR), were developed mostly by Hansch and his coworkers and have been reviewed by Tute (8), Purcell et. al. (9) and Dunn (10). These techniques were utilized by Greenberg (11) in the correlation of odor threshold and suprathreshold data with Log P, the log (n-octanol/water partition coefficient). In the same study it was reported that steric and polar effects as measured by the Taft Steric and Polar Constants poorly correlated with odor intensity data.

The purpose of this paper is to describe how the Quantitative Structure Activity Relationship (QSAR) technique known as the Hansch

Approach was used in deriving mechanistic information about odor intensity as well as insight into how this biological activity may be predicted. This paper will first briefly describe the history of QSAR, the QSAR parameters used, and how substituents for QSAR studies are selected. Several examples of the Hansch Approach used in taste and odor quality studies will next be presented. The balance of the paper will deal with the development of quantitative structure odor intensity relationships which will further expand upon the earlier study reported by this author (11). For example, the use of relatively new QSAR steric parameters in correlations with odor intensity data, and correlations of log P with literature odor intensity data determined on animal panels will be presented. This will be followed by conclusions derived from those studies, and areas of future work.

Historically one of the first QSAR studies was conducted in 1893 by Richet (12) who concluded that the toxicity of ethers, alcohols, aldehydes and ketones was inversely related to their water solubility.

In 1899 Overton (13) and Meyer (14) correlated narcotic activity with lipid solubility (chloroform-water partition coefficients) of a wide variety of non-ionized compounds. They found that narcotic activity increased with increasing lipophilicity until lipid solubility became so high that the substance was virtually water insoluble. They also found that these compounds penetrated tissue cells as though the membranes were lipid in nature. This is the first reported correlation between partition coefficients and biological activity. A second major development in QSAR occurred in 1939 when Ferguson (15) was able to calculate toxic concentrations of a series of compounds from solubility and vapor pressure data.

The next significant advances were made by attempts to use substituent constants rather than physical measurements on the whole molecule. In 1940 Hammett (16) developed the (σ) substituent constants, which measure the degree of electron release/withdrawal of aromatic substituents. Based on the Hammett equation, Hansen (17) correlated bacterial growth inhibition of a series of compounds with their Hammett σ constants.

In the early 1960's Hansch and coworkers developed the Hansch equation. Since then quantum mechanical QSAR and pattern recognition QSAR have emerged. The Hansch approach today is still a widely used technique in medicinal chemistry and insecticide chemistry.

Historically Hansch correlated the Hammett σ constant and log (n-octanol-water partition coefficient) of phenoxyacetic acids with their plant growth regulator activity producing equation 1:

$$\text{Log } A_i = -K_1 (\log P_i)^2 + K_2 \log P_i + K_3 \quad (1)$$

In this equation A_i's represents the activity of the ith member of the series studied and can be in terms of a standard or relative biological response. For comparative purposes A_i is usually the reciprocal of the molar concentration required to elicit a predetermined biological response such as ED_{50}, LD_{50}, etc. The term P_i is the partition coefficient of the compound between the nonpolar biophase of the biological system and its aqueous phase, and accounts for the lipophilic character of the drug, odorant etc.

The K's are constants determined by regression analysis. A detailed derivation of the equation can be found in a review by Tute (8). If activity is a function of the steric and electronic nature of the compound's substituents, these effects are assumed to be included in the term K_3 which can be factored into E_s and σ, the Taft and Hammett constants or other pertinent linear free energy constants (LFER) as shown in equation 2:

$$K_3 = f(E_s, \sigma, \underset{\text{Constant}}{\text{LFER}}) \quad (2)$$

Making the appropriate substitutions produces equation 3:

$$\text{Log } A_i = -K_1 (\log P_i)^2 + K_2 \log P_i + K_3 \sigma + K_4 E_s + \ldots \quad (3)$$

The log P term is the log of the (n-octanol-water partition coefficient). This partition coefficient is used as a reference for the lipophilic character and thus a model for the interaction of compounds with lipoidal biophases. n-Octanol has been most extensively used. In cases where it has been possible to actually measure interactions of drugs with biological phases, the n-octanol-water partition coefficient have been a sufficient model for estimating the interaction. Also much work exists in the literature on this additive constitutive property of organic molecules. In theory, this is a linear free energy substituent constant since the free energy of the partitioning process in the n-octanol-water system is linearly related to that of lipoidal-aqueous biophases.

In conducting a QSAR study using the Hansch Approach substituents must be chosen in order to obtain a wide range of hydrophobicity. This is a necessary requirement in order to determine whether there is an optimum hydrophobicity associated with maximum biological activity. More importantly, the separation of hydrophobic from steric or electronic effects requires that substituents be chosen to prevent colinearity of independent variables. There are several techniques which can be utilized in the rational selection of substituents to meet these requirements. Craig (18) has suggested E_s vs. log P or σ vs. log P plots be constructed to reduce colinearity, but this technique is limited to two independent variables. Hansch, Unger and Forsythe (19) have used cluster analysis as an aid in aromatic substituent selection. Essentially, this is a multidimensional Craig Plot which generates clusters of substituents having similar electronic, hydrophobic, and steric properties. By selecting one or two substituents from each cluster little interrelationship between physical properties can be achieved.

Several examples of the Hansch Approach used in the area of sweetners and odor quality exist in the literature. Hansch and Deutsch (20) found that the relative sweetness of 2-amino-4-nitrobenzenes increased with substituents being more hydrophobic and electron releasing in nature.

Boelens (21) correlated almond odor quality with hydrophobic and steric parameters. Examining Boelens data it was found that a high degree of colinearity between S (Steric parameter) and log P_i existed for the data set of odorants studied. This example illustrates the need for proper substituent selection in order to achieve maximum information from a QSAR study. In the same article a parabolic relationship for musk odor quality was also reported by Boelens with a log Po value (optimum log P value for best odor quality) of 6.24.

Procedures

The procedures for doing the quantitative structure odor intensity relationship study involved the following:

(1)   searching the chemical literature for odor detection threshold values and suprathreshold values of classes of chemical compounds whose members have noncolinear steric, polar and hydrophobic constants,

(2)   calculating or using reported steric (Taft, Charton, Sterimol Constants), Taft polar and hydrophobic (log P) constants, and

(3)   correlating the log of the reciprocal millimolar concentration required for a threshold value or suprathreshold value denoted log (1/c) with the corresponding polar, steric, and log P values.

Classes of chemical compounds having different functional groups and odor descriptors, some of which are useful to the flavor or perfume industries were selected for this initial study. For example, alcohols, aldehydes, pyrazines and various benzenoid compounds which have been isolated in the volatiles of cooked meat as reviewed by Hornstein (22) were studied. For each class of chemical compounds literature threshold values obtained only from one laboratory were used in order to prevent errors associated with technique or methodology between laboratories that occur for threshold determinations as discussed by Guadagni et. al. (2) and Powers and Ware (23).

Suprathreshold odor intensity data from Dravnieks (7) equating odor intensity equivalent to 87 ppm (vol./vol.) of n-butanol was used since it eliminated errors between laboratories which occur for threshold measurements and the n-butanol reference scale has been approved by the ASTM (24) as a standard method of measuring odor intensity.

The log [n-octanol/water partition coefficients] (log P) for compounds selected for this study were obtained from those reported by Hansch et. al. (25), or were calculated from fragmental-constants as reported by Nys and Rekker (26). The Taft Steric (E_s) and polar (σ *) constants were obtained from those values reported by Taft (27).

For alcohols the E_s and σ * values for the substituents bonded to the carbinol moiety were each summed and correlated against log (1/c). For aldehydes and ketones the E_s and σ * values for substituents bonded to the carbonyl group were each summed and correlated against log (1/c). The use of ΣE_s and $\Sigma \sigma$ * has been reviewed by Shorter (28).

In addition to the Taft steric constant, several relatively new steric parameters were used. The υ steric parameter was developed by Charton (29) and is a measure of the degree of branching in substituent groups. The υ parameter for a substituent X is defined as the difference of the van der Waals radii of the X group and hydrogen atom. As in the case of E_s, υ is highly correlated to ester hydrolysis. The υ parameter is much more available and has been measured for a greater range of group type than the E_s constant.

The Sterimol Steric constants were developed by Verloop et. al. (30) to measure steric effects of substituents which are due to a kind of fit to a surface such as when substituents are engulfed in a receptor site. The length (L), minimum width (B$_1$) and maximum width (B$_4$) parameters may provide an improved steric picture over that of parameters such as E$_s$ which are highly correlated to only average radii of substituents.

The use of hydrogen bonding indicator parameter (HB) in Quantitative Structure Activity Relationships has recently been reviewed by Fujita et. al. (31). In that study it was found that an indicator parameter (HB) which represents the "extra" hydrogen-bonding effect on the biological activity is required in the Hansch-type correlations when the relative hydrogen-bonding effect of bioactive compounds on phases involved in the binding at the site of biological action differs from that in the n-octanol-H$_2$O partitioning phases used as the reference to estimate hydrophobicity. Examples were presented in which the HB indicator parameter was used in correlating activity of gaseous anesthetics and the binding of phenyl n-methyl carbamates with acetylcholinesterase inhibition. In this study HB was used to ascertain whether it would improve the correlations involving series of congeners with substituents having appreciable association capability. Non-hydrogen bonders were assigned an HB value of 0 while hydrogen acceptors or donors were each assigned an HB value of 1.

Regression studies of the odor intensity data were carried out using the Continental Can Co. Stepwise Multiple Regression program and the PDP-11-45 mini computer (Digital Equiment Corp.).

Results and Discussion

Results of the regression studies relating literature odor intensity to log P, E$_s$, σ^*, ν and the various sterimol parameters are presented in Tables 1-V. For each equation n is the number of compounds in the data set, R is the correlation coefficient, and S is the equation standard deviation. The numbers in parentheses are the calculated confidence intervals at the 95% level of confidence.

In general it was found from the fifteen sets of data in Tables I-V very good correlations were achieved between log (1/c) and log P or log P and HB with fourteen sets having an equation with a correlation coefficient greater than 0.88 which was at least significant at the 95% level of confidence. Very good correlations with log P were found using literature threshold data as well as suprathreshold data. Thus two different odor intensity parameters correlated well with log P. Odor intensity of homologous series as well as for compounds with different functional groups were found to correlate well with log P, although correlations of the latter were improved by the addition of the hydrogen bonding parameter HB. For example, results in Table IV indicated that log P and HB correlated well with suprathreshold data for a data set of 50 compounds which include hydrocarbons, benzenoids, heteroaromatics, aliphatic ether, ketones, aldehydes, acids and esters.

Specifically odor intensity was poorly related to E$_s$ or Σ E$_s$, and the Sterimol steric parameters for data sets whose hydrophobic and steric

TABLE I: Equations Relating Alcohol Odor Intensity To Hydrophobic, Steric and Electronic Parameters

Intensity Type (Medium)	Eq. No.	Equation	n	R	s	Source
Suprathreshold (Air)	1	$\log (1/c) = 1.50 \pm (0.66) \log P - 1.34 \pm (1.25)$	13	0.83	1.25	A
Suprathreshold (Air)	2	$\log (1/c) = -0.53 \pm (0.71) (\log P)^2 + 3.03 \pm (2.16) \log P - 1.76 \pm (1.18)$	13	0.87	1.16	A
Suprathreshold (Air)	3	$\log (1/c) = 1.40 \pm (0.51) \log P + HB - 2.81 \pm (1.5)$	13	0.88	1.09	A
Suprathreshold (Air)	4	$\log (1/c) = -0.37 \pm (1.44) \ E_s + 0.35 \pm (2.38)$	9	0.22	1.80	A
Suprathreshold (Air)	5	$\log (1/c) = 0.96 \pm (4.91) \quad - 1.14 \pm (3.82)$	9	0.25	1.79	A
Suprathreshold (Air)	6	$\log (1/c) = 0.61 \pm (0.90) \ L -5.62 \pm (8.16)$	9	0.52	1.58	A
Suprathreshold (Air)	7	$\log (1/c) = 0.51 \pm (2.10) \ B_1 -2.22 \pm (1.00)$	9	0.21	1.80	A
Suprathreshold (Air)	8	$\log (1/c) = 0.76 \pm (0.45) \ B_4 -5.00 \pm (3.02)$	9	0.83	1.03	A
Suprathreshold (Air)	9	$\log (1/c) = 0.30 \pm (3.20) \ \sigma* -0.36 \pm (1.88)$	9	0.11	1.35	A
Threshold (Air)	10	$\log (1/c) = 0.78 \pm (0.39) \log P + 3.06 \pm (0.93)$	11	0.83	0.97	B
Threshold (Air)	11	$\log (1/c) = -0.29 \pm (0.15) (\log P)^2 + 1.84 \pm (0.60) \log P + 3.91 \pm (2.01)$	11	0.95	0.55	B
Threshold (Air)	12	$\log (1/c) = -1.27 \pm (1.92) \ E_s + 8.16 \pm (4.43)$	11	0.45	1.56	B
Threshold (Air)	13	$\log 1/c = 0.69 \pm (0.40) \log P + 3.94 \pm (0.92)$	13	0.76	1.05	C
Threshold (Air)	14	$\log (1/c) = -0.33 \pm (0.17) (\log P)^2 + 1.92 \pm (0.63) \log P +3.73 \pm (2.58)$	13	0.93	0.64	C

TABLE I: Equations Relating Alcohol Odor Intensity To Hydrophobic, Steric and Electronic Parameters (Cont'd.)

Intensity Type (Medium)	Eq. No.	Equation	n	R	S	Source
Threshold (Air)	15	$\log (1/c) = -0.21 \pm (1.29)\Sigma E_s + 5.33 \pm (2.92)$	13	0.09	1.60	C
Threshold (Air)	16	$\log (1/c) = 0.81 \pm (6.05)V + 4.31 \pm (4.51)$	13	0.09	1.60	C
Threshold (Air)	17	$\log (1/c) = -1.49 \pm (3.43)B_1 + 10.27 \pm (5.83)$	11	0.31	1.65	C
Threshold (Air)	18	$\log (1/c) = -0.25 \pm (0.69)\Sigma\sigma^* + 5.11 \pm (1.27)$	13	0.06	1.61	C
Threshold (Water)	19	$\log (1/c) = 1.82 \pm (0.40) \log P + 6.08 \pm (0.50)$	6	0.99	0.37	D
Threshold Water	20	$\log (1/c) = -0.29 \pm (0.18) (\log P)^2 + 2.31 \pm (0.36) \log P + 6.48 \pm (0.22)$	6	0.99	0.14	D

A - Dravnieks (7)
B - Laffort (32)
C - Laffort (5)
D - Stahl (33)

properties were not colinear. Poor correlations between log (1/c) and σ * (the Taft polar constant) were observed for the same suprathreshold data sets. This is illustrated in the following examples. Table I indicates that alcohol Suprathreshold odor intensity correlated well with log P and HB; and poorly with ΣE_s or $\Sigma \sigma$ *. The Charton and Sterimol parameters also correlated poorly with odor intensity. (Note that in equation 8, ΣB_4 was colinear with log P resulting in a correlation coefficient of 0.9, thus the importance of B_4 can not be ascertained with this particular data set.)

Threshold data for aliphatic alcohols also correlated well with log P and poorly with ΣE_s and $\Sigma \sigma$ *. Results in equations 11 and 14 indicate a parabolic dependence of alcohol odor intensity upon P. Log Po was found to be 3.17. Thus aliphatic alcohols having a log P value of 3.17 should have maximum odor intensity based upon threshold data. Poor correlations were also found for the Charton and Sterimol parameters. For the threshold data the Sterimol parameters L and B_4 were each highly colinear with log P and thus were not included in Table I.

The alcohol data indicates that the bulkiness of the substituents on the carbinol moiety does not determine the level of odor intensity. The suprathreshold data also indicates that the polar effects of the groups bonded to the carbinol moiety did not effect the level of odor intensity.

Aldehyde and ketone suprathreshold odor intensity correlated well with log P and HB as shown in Table II. No significant relationship between steric or electronic parameters with aldehyde-ketone suprathreshold data was found with the exception of the Sterimol parameter Σ L which was highly correlated to log P (R=0.95). Aldehyde threshold data was found to be linearly related to log P as shown in equations 10 and 13. The same data was poorly correlated with E_s and ν as shown in Table II (eq. 12, 15 and 16). Note that two different aldehyde threshold data sets from two different sources produced very similar equations having slopes, intercepts, correlation coefficients and standard deviations which are not statistically different at the 95% level of confidence (eq. 10 and 13).

This indicates that log P can be used to reproduce predictive equations. The aldehyde-ketone results indicate that the bulkiness of the substituents on the carbonyl group does not determine the level of odor intensity. Suprathreshold correlations indicate that the polar effects of the groups bonded to the carbonyl group does not determine the level of odor intensity. Similar conclusions regarding the importance of hydrophobic and steric effects can be made from the alkane odor intensity -log P and E_s equations in Table III.

The other results in Table III are those of data sets not having noncolinear physical-chemical properties. Log P was highly correlated with these data sets as well. Ethylesters threshold data in air was linearly related to log P (eq. 4) while 3-alkyl-2-methoxy pyrazines had threshold odor intensity which was parabolically related to log P (eq. 7). The pyrazine data indicates that 3-alkyl-2-methoxy pyrazines having a log P value of 2.43 would have the most intense odor of the series.

Benzenoids and heteroaromatics odor intensity was highly correlated to log P. Addition of the HB indicator variable improved this correlation significantly (Table III - eq. 9).

The data in Table IV indicates that odor intensity of a wide variety of

TABLE II. Equations Relating Aldehyde-Ketone Odor Intensity To Hydrophobic, Steric and Electronic Parameters

Intensity Type (Medium)	Eq. No.	Equation	n	R	S	Source
Suprathreshold (Air)	1	$\log (1/c) = 1.49 \pm (1.58) \log P - 0.06 \pm (2.19)$	9	0.64	1.66	A
Suprathreshold (Air)	2	$\log (1/c) = 1.89 \pm (0.66) \log P + 2.23 \pm (0.88) HB - 3.23 \pm (0.89)$	9	0.96	0.66	A
Suprathreshold (Air)	3	$\log (1/c) = -0.47 \pm (1.61) (\log P)^2 + 2.64 \pm (4.33) \log P - 0.44 \pm (1.88)$	9	0.68	1.72	A
Suprathreshold (Air)	4	$\log (1/c) = -1.08 \pm (1.44) \Sigma E_s + 0.07 \pm (1.70)$	7	0.65	1.38	A
Suprathreshold (Air)	5	$\log (1/c) = 2.03 \pm (3.12) \Sigma \nu - 1.73 \pm (4.24)$	7	0.59	1.46	A
Suprathreshold (Air)	6	$\log (1/c) = 1.25 \pm (0.87) \Sigma L - 9.36 \pm (7.16)$	7	0.93	1.02	A
Suprathreshold (Air)	7	$\log (1/c) = -0.43 \pm (7.89) \Sigma B_1 + 2.14 \pm (23.28)$	7	0.06	1.82	A
Suprathreshold (Air)	8	$\log (1/c) = 0.75 \pm (1.98) \Sigma B_4 - 3.01 \pm (10.37)$	7	0.40	1.67	A
Suprathreshold (Air)	9	$\log (1/c) = 0.31 \pm (4.86) \Sigma \sigma^* + 0.63 \pm (1.79)$	7	0.48	1.79	A
Threshold (Water)	10	$\log (1/c) = 0.19 \pm (0.04) \log P + 3.88 \pm (0.24)$	8	0.97	0.21	B
Threshold (Water)	11	$\log (1/c) = 0.15 \pm (0.51) (\log P)^2 + 0.15 \pm (0.15) \log P + 3.81 \pm (0.24)$	8	0.98	0.22	B
Threshold (Water)	12	$\log (1/c) = -3.90 \pm (6.82) E_s + 1.09 \pm (4.33)$	8	0.50	0.79	B
Threshold (Water)	13	$\log (1/c) = 0.42 \pm (0.25) \log P + 3.60 \pm (0.64)$	7	0.97	0.42	C
Threshold (Water)	14	$\log (1/c) = 0.02 \pm (0.11) (\log P)^2 + 0.36 \pm (0.36) \log P + 3.64 (0.25)$	7	0.97	0.22	C
Threshold (Water)	15	$\log (1/c) = -3.24 \pm (6.00) E_s + 3.49 \pm (1.83)$	7	0.53	0.72	C
Threshold (Water)	16	$\log (1/c) = 6.49 \pm (9.91) V + 3.02 \pm (6.79)$	7	0.60	0.54	C

A - Dravnieks (7)

B - Guadagni et. al. (2)

C - Ahmed et. al. (34)

TABLE III. Equations Relating Odor Intensity To Hydrophobic, Steric and Electronic Parameters For Various Compounds

Compounds	Intensity Type (Medium)	Eq. No.	Equation	n	R	s	Source
Alkanes	Threshold (Air)	1	$\log (1/c) = 0.76 \pm (0.42) \log P + 1.20 \pm (1.81)$	7	0.89	0.69	A
Alkanes	Threshold (Air)	2	$\log (1/c) = -0.24 \pm (0.18) (\log P)^2 + 2.57 \pm (1.42) \log P + 1.36 \pm (1.08)$	7	0.97	0.39	A
Alkanes	Threshold (Air)	3	$\log (1/c) = 10.42 \pm (24.0) E_s + 8.12 \pm (1.55)$	7	0.51	0.60	A
Ethylesters	Threshold (Air)	4	$\log (1/c) = 0.42 \pm (0.15) \log P + 5.43 \pm (0.63)$	7	0.96	0.27	A
Ethylesters	Threshold (Air)	5	$\log (1/c) = -0.05 \pm (0.12) (\log P)^2 + 0.45 \pm (0.50) \log P + 5.37 \pm (0.83)$	7	0.96	0.30	A
3-alkyl-2 Methoxy-Pyrazines	Threshold (Water)	6	$\log (1/c) = 2.37 \pm (1.49) \log P + 4.14 \pm (2.16)$	7	0.88	1.46	B
3-alkyl-2-Methoxy-Pyrazines	Threshold (Water)	7	$\log (1/c) = -1.04 \pm (0.77) (\log P)^2 + 5.05 \pm (2.19) \log P + 3.38 \pm (1.23)$	7	0.97	0.77	B
Benzenoids and Heteroaromatics	Threshold (Water)	8	$\log (1/c) = 0.93 \pm (0.61) \log P -0.48 \pm (1.35)$	11	0.75	0.68	C
Benzenoids and Heteroaromatics	Threshold (Water)	9	$\log (1/c) = 1.12 \pm (0.41) \log P + 1.21 \pm (0.72) HB -1.92 \pm (0.86)$	11	0.92	0.43	C

A - Laffort (32)
B - Seifert (35)
C - Stahl (33)

TABLE IV. Equations Relating Hydrophobicity To A Wide Variety of Odorants

Compounds	Intensity Type (Medium)	Eq. No.	Equation	n	R	S	Source
Aldehydes, Ketones	Suprathreshold	1	$\log (1/c) = 0.38 \pm (0.37) \log P + 0.12 \pm (0.45)$	50	0.29	1.75	A
Acids, Esters	(Air)	2	$\log (1/c) = -0.55 \pm (0.22) (\log P)^2 + 2.46 \pm (0.91) \log P -0.85 \pm (0.55)$	50	0.63	1.43	A
Ethers, Alcohols Hydrocarbons, Benzenoids		3	$\log (1/c) = -0.38 \pm (0.19) (\log P)^2 + 2.12 \pm (0.74) \log P + 1.18 (0.48)$ HB $- 5.23 \pm (0.45)$	50	0.80	1.17	A

A - Dravnieks (7)

odorants is related to log P and HB. As before HB improves correlations of log P and odor intensity for nonhomologous series.

Table V presents correlations of animal odor intensity data and log P. Once again log P was highly correlated to odor intensity, for aliphatic acids (2-dog panel), aliphatic alcohols (29 rats) and aliphatic acetates (42 rats).

The concept of odorant hydrophobicity, as measured by log P, determining the level of odor intensity offers insight into the mechanism of olfaction. As discussed by Wright and Burgess (37) it is known from electron microscopy that in vertebrates the olfactory epithelium contains a tangle of cilia floating in a mucus layer. At any instant the cilia which contains many receptor cells may be totally or partially immersed in this mucus layer. Thus the ability of an odorant to partition through the mucus layer and membrane layers of the cilia will affect the concentration of the odorant that reaches the binding sites and thus odor intensity. An odorant may still partition through membrane layers of cilia not in the mucus layer, or membrane layer of receptor cells in the trigeminal nerve, until it reaches the receptor site.

Hansch and Dunn (38) have concluded that the log P coefficient is a measure of the systems sensitivity to hydrophobic effects. Coefficient values greater than 0.85 were typical of hydrophobically sensitive systems such as those found for drugs interacting with membranes. Coefficient values between 0.40 and 0.84 were typical of intermediate hydrophobic sensitivity such as those found for drugs interacting with proteins. All correlations of suprathreshold and threshold data sets had coefficient values typical of those found for drugs interacting with membranes. Aldehyde threshold and alkane threshold produced values typical of intermediate hydrophobic sensitivity. This further supports the view that the partitioning through membrane layers is crucial in determining the odorant concentration at receptor sites and thus odor intensity.

The log P term will also contain a contribution owing to the ability of an odorant to partition from the media in which it is dissolved into the atmosphere. This volatility contribution has been measured by Buttery et. al. (39, 40) and Nawar (41) for compounds in dilute aqueous solutions and is called the air/water partition coefficient (A/W). Table VI presents equations relating log P with log (A/W) for homologous series of methyl ketones, alcohols and aldehydes. For each homologous series log P is linearly related to log A/W. These equations indicate that volatility of odorants in aqueous solutions increases with increasing homolog hydrophobicity. The aldehyde threshold data indicates that the more hydrophobic aldehydes have more intense odors because of their high volatility in aqueous solutions and their ability to partition through biolayers to reach olfactory receptor sites. On the other hand, the 3-alkyl-2-methoxy pyrzaine threshold data (Table III Eq. no. 7) indicates that there is an optimum log P value of 2.43 for maximum odor intensity. This indicates that pyrazines with log P values greater than 2.43 are more volatile in aqueous solutions but have a weaker odor intensity than a pyrazine with a log P value of 2.43; therefore, with a congeneric series the analogs with the highest volatility are not necessarily the most intense odorants.

Similar arguments can be made for alcohol threshold data results.

TABLE V. Equations Relating Animal Odor Intensity Data with Physico - chemical Properties

Compounds	Species	Eq. No.	Equation	n	R	S
Aliphatic Acidss	Canine	1	$\log (1/c) = 1.61 \pm (0.54) \log P + 9.05$	9	0.94	0.77
"	"	2	$\log (1/c) = 0.31 \pm (0.44) (\log P)^2 + 2.26 \pm (1.03) \log P + 9.12$	9	0.96	0.67
Aliphatic Alcohols	Rats	3	$\log (1/c) = 1.25 \pm (0.18) \log P + 5.83$	12	0.98	0.54
"	"	4	$\log (1/c) = -0.11 \pm (0.08) (\log P)^2 + 1.74 \pm (0.36) \log P + 5.70$	12	0.99	0.39
Aliphatic Acetates	Rats	5	$\log (1/c) = 1.11 \pm (0.16) \log P + 2.41$	7	0.99	0.17
"	"	6	$\log (1/c) = -0.15 \pm (0.03) (\log P)^2 + 1.62 \pm (0.12) \log P + 2.15$	7	1.00	0.03

D. G. Moulton and J. T. Eayrs (36)

TABLE VI. Equations Relating Log P with Volatilities of Organic Molecules in Dilute Water Solutions By log Air–Water Partition Coefficients (log A/W)

Compounds	Eq. No.	Equation	n	R	S	Source
Normal Aliphatic Aldehydes	1	$\log P = 3.31 \pm (0.73) \log A/W + 8.11 \pm (1.44)$	6	0.99	0.10	A
Normal Aliphatic Alcohols	2	$\log P = 4.98 \pm (0.59) \log A/W + 17.89 \pm (1.91)$	8	0.96	0.15	B
Methylketones	3	$\log P = 5.50 \pm (1.54) \log A/W + 14.08 \pm (3.43)$	7	0.96	0.30	C

A - Buttery et. al. (39)
B - Buttery et. al. (40)
C - Nawar (41)

For odorants in air this point is further illustrated by considering vapor pressure, log P and HB values and equation no. 2 in Table II. Substitution of log P and HB values for acetone and acetophenone into equation no. 2 in Table II produces log (1/c) values which indicate that over 2,000 times more acetone (vapor pressure = 202 torr at 25°C) is needed to produce the same odor intensity of acetophenone (vapor pressure = 1.09 torr at 25°C) based on molar concentration needed to produce odor intensity equivalent to 87 ppm n-butanol. Thus the more volatile odorant acetone is a weaker odorant in terms of intensity than the more hydrophobic odorant, acetophenone.

The use of log P and HB parameters as a tool for predicting odor intensity seems promising. Although many excellent correlations were obtained as presented in Tables I-V further studies are needed to investigate several unresolved areas. The question on whether log P is linearly or parabolically related to odor intensity for a specific medium needs to be resolved. Six equations in Tables I-V linearly related log P to odor intensity, while five parabolic relationships were observed which had an optimum hydrophobicity (log P) associated with maximum odor intensity. Log Po values observed were 3.17 and 2.90 for alcohols (threshold-air). Alkanes had a log Po value of 5.35 (threshold-air). In aqueous media alcohols had a log Po value of 3.98 while 3-alkyl-2-methoxy pyrazines had a value of 2.43. The animal data indicates that rats had log Po values of 5.40 for acetates and 7.91 for alcohols.

The log Po values differ for the various series. Structurally different sets of compounds acting by the same mechanism on the same receptor sites would all have the same log Po value. Hansch has extensively illustrated this for drugs such as barbiturates having hynotic activity. It is therefore possible that the above compounds interact with different receptor sites and/or by different mechanisms. More work is needed to verify this point.

As discussed by Cammarata and Rogers (42) the more complex the biological system on which a series of bioactive compounds is tested, the more likely the biological activities will be found to be non-linear with respect to partition coefficients. The rationale for this is that compounds with a particular partition coefficient (Po) value achieve sufficient concentrations in a receptor compartment to lead to a maximum in biological response. Compounds with partition coefficients greater or less than Po tend to become involved in kinetic or energetic processes which cause decreased concentrations of the bioactive compound in the receptor compartment. The biological activities of simple test systems may at times show a non-linear dependence with respect to partition coefficients, but this usually occurs when the bioactive substances are intrinsically of high lipophilicity, and a wide range of log P values is represented by the series. It is possible that the observed linear relationships between odor intensity and log P would become parabolic if the authors would have studied data sets with compounds having larger log P ranges such as 5-6.

The equations in Table I indicate that for alcohols odor intensity is parabolically dependent upon log P for threshold values determined in air (Eq. no. 11,14) and in water (Eq. no. 20) and linearly dependent upon log P for suprathreshold values in air (Eq. no. 1). The alcohol odor intensity also could be parabolically dependent upon log P for the suprathreshold values in air, if the authors would have studied additional compounds having log P values of 3.75-5.00 since the log P value that gives optimum odor intensity in equation 11 is 3.17. The log P range for the suprathreshold in air data is (-0.32 to 3.25) which have few data points with log P values greater than the optimum log P value of 3.17. The aldehyde -ketone suprathreshold data also had a narrow log P range of -0.24 to 2.75. The same can be said for aliphatic aldehydes, esters and benzenoid threshold data sets log P ranges. More work is needed in this area.

Another area of further study is the reproducibility and accuracy of derived predictive equations. Two different data sets of aliphatic aldehyde threshold values in water were subjected to QSAR techniques to determine whether log P can be used to accurately reproduce predictive equations for odor intensity data of a compound in a given medium determined by two different laboratories. Results in Table II indicate that equations 10 and 13 have slopes, intercepts, correlation coefficients and standard deviations which are not statistically different at the 95% level of confidence. Both data sets also produced equations giving poor correlations of E_s and log (1/c) which were not statistically significant.

Summary

The use of the QSAR technique known as the Hansch Approach in the investigation of odor intensity and odorant physico-chemical properties has indicated that hydrophobic properties of homologous series of compounds, not steric or polar properties, are highly correlated to the level of odor intensity. This was shown to be the case for literature odor threshold and suprathreshold data determined at different laboratories using various media. The poor correlation between odor intensity and the steric properties of molecules (Taft Steric Constant) which had been reported earlier by this author (11) have been further verified by the use of Charton and Verloop Sterimol steric parameters.

The hydrophobicity term as measured by log P, the log n-octanol/water partition coefficient , indicates that the ability of an odorant to partition from the medium in which it is dissolved into the atmosphere and its ability to partition through mucus and membrane layers to reach olfactory receptor sites is highly correlated to odor intensity. Results of this study also indicated that within a congeneric series, the analogs with the highest volatilities are not necessarily the most intense odorants.

The ability of these techniques to predict odor intensity of organic compounds in a given medium seems promising. Many good correlations between literature odor intensity data and log P were observed for different media and for two different methods of measuring odor intensity, odor threshold and suprathreshold techniques. Log P correlated well with homologous and nonhomologous series. The addition of a hydrogen bonding indicator parameter, HB, to equations relating odor intensity to log P for nonhomologous series of compounds resulted in significantly improved correlations in four cases. The reproducibility of the predictive power of the derived equations was shown to be very good. This was demonstrated by predictive equations for literature aldehyde threshold values determined in water by two different laboratories. The derived equations were shown to be statistically equivalent at the 95% level of confidence.

Further work is needed in this area before a general predictive equation can be derived relating odor intensity of compounds in a given media to log P and HB. The question on whether log P is linearly or parabolically related to odor intensity needs to be resolved. Data sets of odorants having large log P ranges of 5-6 need to be studied to resolve this issue.

Hopefully, further evaluations of log P as an odor intensity predicting tool will generate general equations relating log P to odor intensity for a wide range of important flavor compounds in specific media. A study relating taste intensity and physico-chemical properties of organic molecules is presently being conducted in this laboratory. The taste intensity study may provide information which when used in conjunction with odor intensity equations may aid synthetic organic chemists in designing novel flavor compounds with optimum flavor intensities.

Literature Cited

1.   Davies, J. T.; Taylor, F. H. Biol. Bull. Woods Hole, 1959, 117, 222.
2.   Guadagni, D. G.; Buttery, R. G.; Okano, S. J. Sci. Fd. Agric., 1963, 14, 761.
3.   Beck, L. H. Ann. N.Y. Acad. Sci., 1964, 116, 228.
4.   Laffort, P. in "Olfaction and Taste III", Pfaffmann, C. Ed., Rockefeller University Press, New York, NY, (1969); pp. 150-151.
5.   Laffort, P.; Patte, F.; Etcheto, M. Ann. N.Y. Acad. Sci., 1974, 237, 192.
6.   Dravnieks, A. Ann. N.Y. Acad. Sci., 1974, 237, 144.
7.   Dravnieks, A. in "Flavor Quality: Objective Measurement", Scanlan, R. A. Ed., ACS Symposium Series 51, American Chemical Society, Washington, DC, 1977, pp. 11-28.
8.   Tute, M. S. Advances in Drug Research, 1971, 6, 1.
9.   Purcell, W. P.; Bass, G. E.; Clayton, J. M., "Strategy of Drug Design: A Guide to Biological Activity", John Wiley and Sons, New York, NY, 1973.
10. Dunn, W. J. Ann. Reports in Med. Chem., 1973, 8, 313.
11. Greenberg, M. J. J. Agric. Food Chem., 1979, 27, 347.
12. Richet, M. C. C.R. Soc. Biol., 1893, 45, 75.
13. Overton, E. Z. Physikol. Chem., 1897, 22, 189.
14. Meyer, H. Arch. Exptl. Pathol. Pharmakol., 1899, 42, 109.
15. Ferguson, J. Proc. Roy. Soc., Ser. B., 1939, 127, 387.
16. Hammett, L. P., "Physical Organic Chemistry", 1st. ed., McGraw-Hill, NY, 1940; pp. 184-199.
17. Hansen, O. R. Acta Chem. Scand., 1962, 16, 1593.
18. Craig, P. N. J. Med. Chem., 1971, 14, 680.
19. Hansch, C.; Unger, S. H.; Forsythe, A. B. J. Med. Chem., 1973, 16, 1217.
20. Hansch, C.; Deutsch, E. W. Nature, 1966, 211, 75.
21. Boelens, H. in "Structure Activity Relationships in Chemoreception, Benz, G. Ed., Information Retrieval Limited, London, 1976; pp. 197-206.

22. Hornstein, I. in "The Science of Meat and Meat Products", Price, J. F. Schweigert, B. S. Ed., W. H. Freeman, San Francisco, CA., 1971; pp. 348-363.
23. Powers, J. J.; Ware, G. O. Chem. Senses and Flavor, 1976, 2, 241.
24. ASTM E544, "Recommended Practice for Odor Suprathreshold Intensity Referencing", Am. Soc. Test. Materials, Philadelphia, PA (1975).
25. Hansch, C.; Leo, A.; Elkins, D. Chem. Rev., 1971, 71, 525.
26. Nys, G. G.; Rekker, R. F. Eur. J. Med. Chem., 1974, 9, 361.
27. Taft, R. W. in "Steric Effects in Organic Chemistry, Newman, M. S. Ed., Wiley, New York, NY 1956; pg. 556.
28. Shorter, J. in "Advances in Linear Free Energy Relationships", Chapman, N. B., Shorter, J. Ed., Plenum Press, New York, NY, 1972; pp. 71-119.
29. Charton, M. J. Amer. Chem. Soc., 1975, 97, 1552.
30. Verloop, A.; Hoogenstraaten, W.; Tipker, J. "Drug Design", Vol. VII, Ariens; E. J. Ed., Academic Press: New York, 1976; p. 165.
31. Fujita, T.; Nishioka, T.; Nakajima, M. J. Med. Chem., 1977, 20, 1071.
32. Laffort, P. Comptes. Rendus des Seances de la Societe de Biologie, 1968, 162, 1704.
33. ASTM DS48, "Compilation of Odor and Taste Threshold Values Data", Stahl, W. H. Ed., Am. Soc. Test. Materials, Philadelphia, PA (1973).
34. Ahmed, E. M.; Dennison, R. A.; Dougherty, R. H., Shaw, P. E. J. Agric. Food Chem., 1978, 26, 187.
35. Seifert, R. M., Buttery, R. G.; Guadagni, D. G.; Black, D. R.; Harris, J. G. J. Agric. Food Chem., 1970, 18, 246.
36. Moulton, D. G.; Eayrs, J. T. Animal Behav., 1960, 8, 117.
37. Wright, R. H.; Burgess, R. E. Method Chemicum, 1977, 11, 268.
38. Hansch, C.; Dunn, III, W. J. J. Pharm. Sci., 1972, 61, 1.
39. Buttery, R. G.; King, L. C.; Guadagni, D. G. J. Agric. Food Chem., 1969, 17, 385.
40. Buttery, R. G.; Bomben, J. L.; Guadagni, D. G.; King, L. C. J. Agric. Food Chem., 1971, 19, 1045.
41. Nawar, W. W. J. Agric. Food Chem., 1971, 10, 1057.
42. Cammarata, A.; Rogers, K. S., in "Advances in Linear Free Energy Relationships", Chapman, N.B., Shorter, J. Ed., Plenum Press, New York, NY, 1972; pp. 420-435.

RECEIVED November 18, 1980.

Odorants as Chemical Messengers

JOHN N. LABOWS, JR.

Monell Chemical Senses Center, 3500 Market Street, Philadelphia, PA 19104

The relationship between chemical structure and perceived odor has been studied by electrophysiological, chemical-analytical, and psychophysical techniques. Certain odorants in addition to being detected by the olfactory system evoke specific behavioral responses. Recent studies on various mammalian species have attempted to equate specific odor sources with behavioral patterns and to profile the odorants in hopes of identifying the biologically active components (1). In addition, studies on human odor suggest similarities in odor sources and types with other mammalian species and also suggest some of these odors may be reflective of internal body processes.

Our initial research efforts have been directed at chemically characterizing the odors which normally emanate from the body and using this information to diagnose disease states, sexual receptivity, and stress. Vaginal secretions, saliva, secretion from the apocrine gland in the axillae (underarm), and sebum from the sebaceous gland all represent unique substrates which can be metabolized by the resident microorganisms to generate odoriferous materials. Table I summarizes the useful types of information which may be contained in these odors. Described below are some of the attempts at profiling these odors and relating them to physiological states.

The present interest in the characterization of both animal and human secretions has paralleled the development in psychophysical measurement techniques and in analytical methods such as headspace concentration, gas chromatography, and the combination of gas chromatography/mass spectrometry (GC/MS) which have made it possible to routinely separate and identify submicrogram quantities of organic compounds. GC/MS profiling of the small organic compounds present in body secretions, such as blood serum, cerebrospinal fluid, and urine of diseased and healthy individuals, has provided useful diagnostic information (2). The metabolic profiles are analyzed for qualitative or quantitative changes in individual components which might correlate with the onset of disease processes or the female reproductive cycle.

0097-6156/81/0148-0195$05.00/0

GC/olfactory analysis is useful for determining which components
of these complex mixtures contribute to the observed odor (3).

Table I   Diagnostic Potential of Human Odors

Odor Source	Information Content	Microorganisms Acting on
Scalp	-----	Sebum
Oral	Time of Ovulation Periodontal Disease Gastrointestinal Disorders	Saliva
Axillae	Stress Level Mental Health	Apocrine Secretion
Vaginal	Time of Ovulation Metabolic Infertility	Vaginal Secretions
Foot	Bacterial Infection	Eccrine Sweat Epidermal Lipid

Odor and Disease:

   Systemic disease processes such as gastrointestinal disorders
and diabetic keto-acidosis (acetone) manifest themselves in
odors associated with breath and/or saliva (4). The classic
uremic breath odor has been described as 'fishy' or 'ammoniacal'
and involves the presence of dimethylamine and trimethylamine in
the breath (5). Elevated levels of mercaptans and C_2-C_5
aliphatic acids are found in the breath of patients with cirrhosis
of the liver (6). Other illnesses such as skin ulcers, gout,
typhoid, diphtheria, smallpox and scurvy have been reported to
have distinct odors (7). In most cases no odor description or
chemical characterization of the odor has been attempted.

   The most important use of body odors in disease diagnosis
relates to the infant diseases involving errors in amino acid
metabolism. Strong and unusual odors are manifest in the breath,
sweat, and urine of these individuals. Table II summarizes
several known acidurias, the amino acids that are not properly
metabolized, and the odors associated with the compounds which
accumulate and can be detected in the urine (8). In the case of
the Maple Syrup Urine and Oasthouse syndrome, the keto- and
hydroxy- acids which have been identified may not be responsible
for the observed maple and celery/yeast odors (9). Alternatively,
these odors could be the result of conversion of 2-keto-butyric
acid to methyl-ethyl-tetronic acid (Slusser's lactone) which is
used as an extender in maple and celery flavors and has a maple
syrup-like odor (R. Soukup, personal communication). With these
acidurias it is imperative that an immediate diagnosis is made,
since corrective diet can prevent the brain damage that results
from these diseases. This is readily done on an olfactory basis
which can subsequently be supported by gas chromatographic

Table II Metabolic Disorders in Infants (63)

	Amino Acid(s)	Enzyme Defect	Compound(s) Accumulated
MSUD	leucine, valine isoleucine	branched chain decarboxylase	2-hydroxy acids 2-keto acids (maple-syrup)
Oasthouse	methionine	methionine utilization	2-keto-butyric acid 2-hydroxy-butyric acid (celery, yeast)
PKU	phenyl alanine	phenyl alanine hydroxylase	phenyl pyruvate phenyl acetic acid (mousy, horsey)
Sweaty Feet	leucine	isovaleryl CoA dehydrogenase	isovaleric acid (sweaty)

analysis of the urine. It is accepted procedure for the
pediatrician to 'smell his patient' and at least one medical
school uses odors as a part of its lecture material (10). As we
understand what odors are associated with various disease processes,
it would be appropriate for the physician to use olfaction for a
diagnosis.

There appears to be a relationship between various oral
pathologies and the chemicals found in human saliva (11).
Various volatile compounds such as skatole, indole, sulfides, and
long chain alcohols have been identified in the headspace of
saliva samples. These materials increase in both a quantitative
and qualitative fashion with varying degrees of periodontitis.
Specifically, alkyl pyridines appear to be present in the saliva
only in individuals with periodontal disease. The monitoring of
these compounds may allow the detection of the early stages of
this disease process which effects 60–70% of the population.

Odor and Communication: Mammalian

Studies which have been undertaken to implicate specific
chemicals in mammalian olfactory signals must first be considered
in order to appreciate the possibility of human odor communication.
Chemical odorants, present in animal skin glands, urine, saliva,
and vaginal fluids have pronounced physiological and behavioral
effects (1,12). The scent-marking skin glands are either
apocrine-like and analogous to the human apocrine gland or are a
combination of apocrine-sebaceous glands. A variety of these
glands present in the rabbit and deer convey alarm and fright
messages as well as information on individual identity (13).
The isolated boar ketones, 5α-androst-16-en-3α-ol (androstenol)
and androst-16-en-3-one (androstenone) secreted by the sub-
maxillary gland, have a direct effect on the sexual receptivity
of the sow and are used commercially to assist in artificial
insemination (14). The fact that estrus can be determined in
the sow by her response to these compounds suggests that there is
a heightened acuity for these compounds at the time of ovulation.
This is similar to the increase in olfactory acuity for certain
compounds noted in human females prior to ovulation (15). A
somewhat unique but analogous situation is the elephant temporal
gland which is an apocrine gland that is active under stress and
possesses an 'elephanty odor' (16). Table III summarizes some of
the mammalian communication systems that have been studied and
the chemicals which have been found to have behavioral effects.
In some cases there are unique odors, such as 'rabbit odor',
'monkey odor', 'deer odor', which are associated with specialized
skin glands and specific chemical structures (13).

The characterization of a behaviorally active chemical is a
tedious task and involves isolation and structural identification
of numerous constituents from a secretion. A suitable bioassay,
which involves presenting the chemical(s) to the animal in a

Table III   Mammalian Chemical Communication

Animal	Chemical (source)	Behavior	Reference
Boar	androstenol and androstenone (submaxillary gland)	induces lordosis	14
Deer	Cis-4-hydroxydodec-6-enoic acid lactone (urine)	sniffing licking	17
Marmoset monkey	butyrate esters of long-chain alcohols (circumgenital gland)	subspecies identity	18
Hamster	dimethyl disulfide (vagina)	elicits attraction	19
Rhesus monkey	short-chain aliphatic acids (vagina)	sexual activity not reproducible	20, 21
Dog	methyl-p-hydroxybenzoate (vagina)	sexual activity	22
Rabbit	cis-undec-4-enal (anal gland)	'rabbity odor' heart rate	23
Reindeer	short-chain acids; ketones (interdigital)	sniffing	24, 25
Pronghorn	isovaleric acid (subauricular)	-----	26
Reindeer	aldehyde, alcohols (tarsal)	-----	27

natural context and 'measuring' a behavioral response, is then
necessary to determine if the chemical(s) are of interest to the
animal (1,17,28). Though many mammalian secretions have been
found which give behavioral responses, few chemicals with
definitive effects have been characterized (Table III). The best
example is the two androgen steroids used by the boar. Recently,
methyl-p-hydroxybenzoate has been isolated from the vaginal
secretions of female dogs and shown to be a highly effective
sexual excitant to males (22).

An understanding of the chemical language that controls
the social and feeding behavior of an individual species would be
useful in the care and breeding of that species. However,
research on the assignment of structure-activity relationships in
mammalian behavior is only in its infancy.

Odor and Communication: Human

Anecdotal stories in the literature refer to the ability of
human odors to effect sexual and social behavior (29).
Psychologists have recently attempted to decipher the information
content of these various odors (30). The control of endocrine
states by odor is suggested by the work on the synchronization
of cycles of females living together (31). Russell demonstrated
that subjects can detect sexual differences and individual
identity using axillary odors (32). In a similar experiment using
a different protocol, Doty determined that individuals equated
the more intense odors with male subjects (20). No effects were
found for aliphatic acids on sexual behavior (33), while neither
the acids nor androstenol had any significant effects on individual judgments (34). In the presence of the androstenol,
photographs of women were judged as more attractive although no
control odors or other 'synthetic musks' were evaluated in this
study (35).

The chemical and psychological changes associated with the
menstrual cycle include changes in olfactory acuity as well as
cyclic changes in numerous biochemical processes. The latter may
be reflected in cyclical variations in body odors, as is the
case in many mammalian species where information on female
receptivity is transmitted to the male through odors from body
secretions. Odors from the mouth and vagina have been examined
as possible sources of chemicals which undergo cyclical changes.
Preliminary work with female breath samples has centered on
three volatile sulfur compounds (hydrogen sulfide, methyl
mercaptan and dimethyl sulfide) which are primarily responsible
for endogenous bad breath ("halitosis"). These three compounds
were found to change in cyclical fashion increasing at the time
of ovulation and again during menstruation (36). With a gas
chromatograph adapted for the detection of sulfur compounds,
these materials can be quantitated at the low nanogram levels.
Their increase corresponds to increases in both bacterial counts
and in exfoliation of cells in the oral cavity.

Olfactory analysis of vaginal odors has shown that human
observers rate the odor least unpleasant and less intense at
the time of ovulation. However, the large variations in response
on individual subjects suggests that this is not a useful predictive approach (37). Detailed chemical profiling of vaginal
secretions has led to the identification of a variety of low
molecular weight organic compounds (38). Long chain acids and
alcohols, 3-hydroxy-2-butanone, dimethysulfone, furfural, cresol,
phenol, furfuryl alcohol, pyridine, propylene glycol, glycerol,
benzoic acid, and cholesterol were consistently present in all
subjects. Lactic acid concentrations did rise at midcycle and
this information may be useful in predicting the time of ovulation.
Short-chain aliphatic acids were present in only six of fourteen
women and did not vary in concentration in a cyclical manner.
These aliphatic acids, referred to as copulins, were found
originally in rhesus monkey vaginal secretions; but their
pheromonal effects have been questioned (20,21).

The odorants that may be of importance for human olfactory
communication are those for which man possesses specific
olfactory receptors as shown by the studies on specific anosmias,
i.e. the inability to detect the odor of specific chemicals.
These odors include spermous, musky, fishy, urinous, malty and
sweaty, and can be related to some observed human odors (39).
Thus, Amoore suggests that, if we have a specific olfactory
receptor for a given odorant then that odorant might be naturally
given off by the body. The sweaty odor of isovaleric acid is
probably part of the foot odor and is produced by the action of
skin bacteria on apocrine secretion (see below). Pyrolline, the
spermous odor, has been shown to be produced by enzymatic break-
down of the polyamines in semen (40). Androst-16-en-3-one, the
urinous primary odor, has axillary-like odor; the related
androstenol, which is found in urine, is perceived as a musky odor
to some individuals (41). Both steroids are found in axillary
sweat and may be formed as metabolites of apocrine secretion.
Chemicals which fit the malty anosmia have not as yet been
reported from human odor sources. The natural musks, such as
cycloheptadecenone (civet), were first obtained from animal
scent glands.

It is of interest to note that observed odorants are in most
cases metabolic by-products of human secretions, rather than
odorants which were directly secreted. The same situation may be
true in a number of mammalian species where bacteria may be
involved in the eventual formation of chemicals used in odor
communication (42).

Odor Analysis:

There has been an interest in developing techniques for
sampling total body odors. Dravnieks' group has sampled the
volatiles emitted by individuals by placing them in a glass
cylinder and sweeping the tube with air to concentrate the

volatiles (3). He has also developed systems for sampling skin
and axillary odors. Ellin used a telephone booth-like chamber in
which human volatiles were sampled (43). Here approximately
300-400 individual chemicals were detected and 135 identified.
The object of these trials was to explore the possibility that
body odors might be unique to a given individual or a given race
and serve as a personal signature. Room air also has been
sampled in the presence and absence of individuals in an attempt
to determine what contaminants were added to the environment by
body volatiles (44). This is particularly relevant to restricted
environments such as submarines and space cabins were air re-
circulation is a necessity. The thermally induced total body
sweat of schizophrenic patients was collected for analysis of
unique odors by the use of large plastic bags (45). In all of
these collections, including one on axillary odor using cotton
pads (46), no volatile chemicals which represent specific 'human
odors' were identified. The sampling and identification of
'body odors' and their role in diagnosing and monitoring disease
states has recently been reviewed (47).

A major contributor to whole body odor are the skin odors
which result from the interaction of microorganisms with
secretions from the eccrine, sebaceous and apocrine glands.
These secretions differ in their chemical composition and thus
provide unique substrates for the organisms (48). The eccrine
glands which are present over most of the body are the thermo-
regulatory sweat glands which respond to physical activity. The
eccrine secretion has been well characterized and consists of an
aqueous solution of inorganic salts and amino acids which has no
significant odor. The sebaceous glands which are located
primarily on the forehead, face and scalp are under hormonal
control and secrete lipid materials such as triglycerides,
cholesterol and wax esters. This secretion has a slight pleasant
odor but can be readily metabolized by skin microorganisms. The
third glandular system is that of the apocrine glands which are
located primarily in the genital and axillary areas. They become
active at puberty because of the presence of androgen steroids
from the adrenal glands, testes and ovaries, and secrete in
response to emotional situations (49). Analysis has shown that
the secretion contains protein (10%), cholesterol (1%), and
steroids (~0.02%).

The most productive approach to the study of skin odors
has been the duplication of the natural odors in vitro by incubating the normal skin microorganisms with these secretions.
For example, the yeast Pityrosporum ovale, the major scalp
resident, is able to metabolize lipid substrates to 4-hydroxy-
acids which readily undergo ring closure to the volatile and
odorous γ-lactones. The technique of headspace concentration on
Tenax followed by gas chromatographic/mass spectrometric analysis
has been used to profile all the volatiles produced by Pityrosporum (Figure 1). These compounds include isopentanol, benzyl

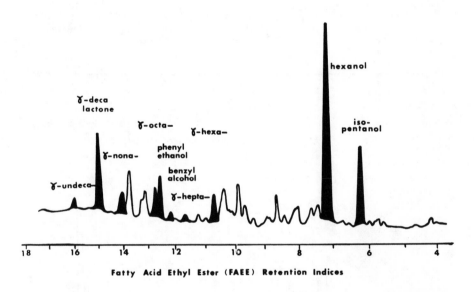

Figure 1.   Odor profile of Pityrosporum ovale *(50)*

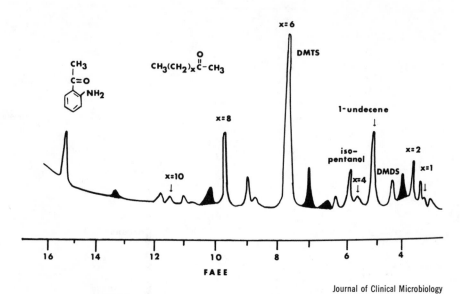

Journal of Clinical Microbiology

Figure 2.   Odor profile of Pseudomonas aeruginosa *(51a)*

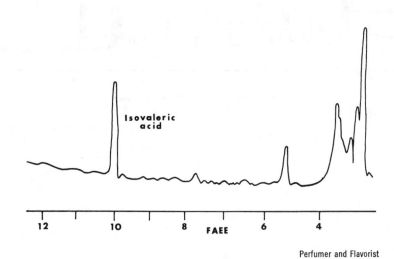

Perfumer and Flavorist

Figure 3.   Micrococci on apocrine secretion (63)

alcohol, phenyl ethanol, and several lactones, including
γ-octalactone (coconut flavor); γ-nonalactone (cream, fruity);
γ-decalactone (peach, pear). The odor of the culture is similar
to that of unwashed hair and closely matches that of γ-decalactone, the major lactone component. Because of the compounds it
produces, this scalp microorganism has the potential to be used
for the natural formation of flavor additives (50). Interestingly,
a similar lactone-profile is observed for another yeast.,
Sporobolomyces odorus isolated from orange leaves (51).

This odor-profile may also be used for the detection of the
Pityrosporum genus, since other yeasts that may be found as
transients on the skin grown on the same media failed to yield
any lactones. In addition, when sebum is the major substrate
and longer incubation times are used a 'scalp odor' is generated.
Our preliminary headspace analysis suggestes that this odor
contains short-chain aliphatic acids in addition to the lactones.
The formation of odors on the scalp may be a cooperative effort
of Propionibacterium acnes, which readily hydrolyzes
triglycerides, and Pityrosporum ovale, which can metabolize the
resultant fatty acids and/or glycerol to various odorants.

The odor profile of a skin pathogen, Pseudomonas aeruginosa
has also been investigated. This organism, though present
normally in some individuals, is responsible for serious infections in burn victims and in lung infections in cystic
fibrosis patients. The volatile profile is shown in Figure 2
where 2-aminoacetophenone (2-AA), methyl ketones and sulfide are
the major unique odorants (51a). The 2-AA, which imparts a
grape-like odor to the culture, is formed from tryptophan and is
characteristic of this species (52). The methyl ketones also
appear to be species specific and may be of value in the detection
of lung infections through breath analysis.

The unique human axillary odor is the result of microbial
action on an odorless secretion. The two major residents of the
axillae are diphtheroids (lipophilic and large colony) and the
micrococci bacteria. Specific odorants can be produced by
incubating these bacteria with apocrine secretion either on a
cleansed forearm or in a test tube. The micrococci produce a
sweaty, acid odor which by headspace analysis has been shown to
be isovaleric acid (Figure 3). The diptheroids also produce this
acid, but its odor is masked by other odor components which
impart a heavier 'apocrine odor' to the incubated sample (63).
Bacterial sampling along with olfactory analysis of individual
subjects further demonstrates that the 'apocrine odor' is
associated with the diphtheroid bacteria. These odorants, which
represent unique human odors in analogy to the animal scents,
are presently being investigated. However, the following
experimental observations relate to the possible identification
of these odorants. The boar pheromone, androst-16-en-3-one and
its precursor, androsta-di-4,16-en-3-one, both have intense
odors which closely resemble the 'apocrine odor' (53). Both of

Table IV.   Steroids Found in Human Axillae *(63)*

Steroid	Sample	Reference
Androst–4–ene–3, 17–dione Androsterone (sulphate) DHA (sulphate) Cholesterol	Axillary Hairs and sweat	55
Androst–4–ene–3, 17–dione Pregn–5–en–3β–ol–20–one	Axillary Sweat	56
5α–Androst–16–en–3α–ol	Axillary Sweat	57
5α–Androst–16–en–3–one	Axillary Sweat	58,58a,59
Androsterone (sulphate) DHA (sulphate) Cholesterol	Apocrine Secretion	60

these 16-androstene steroids in addition to 5α-16-androsten-3α-ol circulate in human blood (54). Trace amounts of androstenone and androstenol as well as other steroids have been reported to be present in human axillary sweat (Table IV). More recently we have found that heated apocrine secretion (>150°) gives an apocrine-like odor. The major contributors to this odor are isomeric androstadien-17-ones and androst-2-en-17-one which arise from the thermal breakdown of dehydroepiandrosterone and androsterone sulfates respectively (60). Thus the apocrine secretion contains specific steroid materials, in addition to cholesterol, which may be metabolized to the odorous Δ^{16}-androgens by the diphtheroid bacteria. Whether this in fact occurs remains to be demonstrated. However, if it is the case, the fact that about 50% of the population is anosmic to these odorants (61) suggests that axillary odor would be perceived as a sweaty odor by anosmic individuals, whereas others would perceive an 'apocrine odor'. Finally, the fact that these steroids have demonstrated sexual effects in one animal suggests that they might also be physiologically active in other species.

The apocrine secretion and the resultant odor is a normal response to emotional stimuli. Dehydroepiandrosterone, which is present in the apocrine secretion, also has been reported to increase in urine in individuals under stress (62). Thus, a sensitive method for monitoring of the activity of the apocrine gland could provide information relative to the emotional state of an individual.

Since man possess odor sources similar to mammalian species, it is of value to determine both the nature and the biochemical origin of these odorants. Profiling of human odors represents a non-invasive technique which might prove useful in the detection of many metabolic and infectious disorders and for monitoring normal body processes. Alternatively, we may be unknowingly emitting and perceiving odorants which could effect our interpersonal relationships. Only further research in this area will determine to what extent this occurs.

Acknowledgements

The work described on skin odors is in cooperation with the Department of Dermatology, University of Pennsylvania and is supported by a grant from the National Institute of Arthritis, Metabolic, and Digestive Diseases.

Literature Cited

1a. Müller-Schwarze, D., Mozell, M.M., Eds. Chemical Signals
      in Vertebrates; Plenum Press, N.Y., 1977.
1b. Müller-Schwarze,D., Silverstein, R.M., eds. Chemical
      Signals Vertebrates and Aquatic Invertebrates; Plenum
      Press, N.Y., 1980.
2.   Jellum, E., J. of Chromatogr., 1977 143: 427.
3.   Dravnieks, A., J. Soc. Cosmet. Chem., 1975 26: 551.
4.   Krotoszynski, B., Gabriel, G. O'Neill, H., J. Chrom. Sci.,
      1977, 15: 239.
5.   Simenhoff, M.L., Burke, J.F., Saukkonen, J.J., Ordinario,
      A.T., Doty, R., New Engl. J. Med. 1977, 297: 132-135
6.   Chen, S., Mahadevan, V., Zieve, L., J. Lab. Clin. Med.,
      1970, 75: 622-628.
7.   Lidell, K., Postgrad. Med. J., 1976, 52: 136.
8.   Mace, J.W., Goodman, S.I., Centerwall, W.R., Chinnock, R.F.,
      Clin. Ped., 1976, 15: 57.
9.   Sulser, H., Depizzol, J., Buchi, W., J. of Food Sci., 1967,
      32: 611-615.
10. Lukas, T., Berner, E., Kanakis, C., J. Med. Educ., 1977,
      52: 349.
11. Kostelc, J.G., Preti, G., Zelson, P.R., Stoller, N.H.,
      Tonzetich, J., J. Periodont. Res., 1980 15: 185-192.
12. Freeman, S.K., J. Soc. Cosmet. Chem., 1978, 29: 47.
13. Mykytowycz, R., Goodrich, B.S., J. of Inv. Derm., 1974,
      62: 124.
14. Reed, H., Melrose, D., Patterson, R.L.S., Brit. Vet. J.,
      1974, 130: 61-67.
15. Doty, R.L., Ed., Mammalian Olfaction, Reproductive
      Processes, and Behavior, Academic Press, 1976.
16. Adams, J., Garcia, A., Foote, C.S., J. Chem. Ecol., 1978,
      4: 17.
17. Müller-Schwarze, D., Ref. 1, p. 413.
18. Epple, G., Golob, N.F., Smith, A.B., in Chemical Ecology:
      Odour Communication in Animals, ed. F.J. Ritter, Elsevier/
      North Holland Biomedical Press, 1979 p. 117-130.
19. Singer, A.G., Agosta, W.C., O'Connell, R.J., Pfaffmann, C.,
      Bowen, D.V., Field, F.H., Science, 1975, 191: 948-950.
20. Goldfoot, D.A., Kravetz, M.A., Goy, R.W., Freeman, S.K.,
      Hormones and Behavior, 1976, 7: 1-7.
21. Michael, R.P., Bonsall, R.W., Zumpe, D., Evidence for
      Chemical Communication in Primates, in Vitamins and
      Hormones, Academic Press, 1976, 137-183.
22. Goodwin, M., Gooding, K.M., Regnier, F., Science, 1979,
      203: 559.
23. Goodrich, B.S., Hesterman, E.R., Murray, K.E., Mykytowycz, R.
      Stanley, G., Sugowdz, G., J. Chem. Ecol., 1978, 4(5),
      581-594.
24. Anderson, G., Brundin, A., Anderson, K., J. of Chem. Ecol.
      1978, 5(3): 321-333.

25.   Brundin, A., Anderson, G., Anderson, K., Mossing, T.,
        Kallquist, L., J. Chem. Ecol., 1978, 4: 613-622.
26.   Müller-Schwarze, D., Müller-Schwarze, C., Singer, A.G.,
        Silverstein, R.M., Science, 1974, 183: 860-862.
27.   Anderson, G., Anderson, K., Brundin, A., Rappe, C., J. Chem.
        Ecol, 1975, 1(2): 275-281.
28.   Beauchamp, G.K., Doty, R.L., Moulton, D.G., Mugford, R.A.,
        in Mammalian Olfaction, Reproductive Processes, and
        Behavior ed. Doty, Academic Press, NY, 1976, 144-157.
29.   Bloch, I., Odoratus Sexualis, Panurge Press, New York, 1934.
30.   Doty, R.L., Ref. 1, p. 273.
31.   McClintock, M.K., Nature, 1971, 229: 244.
32.   Russell, M.J., Nature, 1976, 260: 520.
33.   Morris, N., Udry, R., J. Biosoc. Sci., 1978, 10: 147-157.
34.   Cowley, J., Johnson, A., Brooksbank, B.,
        Psychoneuroendocrinology, 1977, 2: 159-172.
35.   Kirk-Smith, M., Booth, D., Carroll, D., Davies, P.,
        Res. Comm. in Psychol., Psychiatry, and Behavior, 1978,
        3: 379-384.
36.   Tonzetich, J., Preti, G., Huggins, G.R., J. Int. Med.
        Res., 1978, 6: 245.
37.   Doty, R., Ford, M., Preti, G., Huggins, G., Science, 1975
        190: 1316-1318.
38.   Preti, G., Huggins, G.R., in The Human Vagina, ed. E. Hafez
        and T. Evans, Elsevier/North Holland Biomedical Press,
        1978.
39.   Amoore, J., Chemical Senses and Flavor, 1977, 2: 267.
40.   Chvapil, M., Eskelson, C., Jacobs, S., Chvapil, T., Russell,
        D.H., Obstet. Gynecol., 1978, 52: 88.
41.   Kingsbury, A.E., Brooksbank, B.W.L., Hormone Res., 1978,
        9: 254-270.
42.   Albone, E., Gosden, P., Ware, G., Ref. 1a, p. 35.
43.   Ellin, R.L., Farrand, R.L., Oberst, F.W., Crouse, C.L.,
        Billups, N.B., Koon, W.S., Musselman, N.P., Sidell, F.R.,
        J. Chromatogr., 1974, 100: 137.
44.   Savina, V.P., Sokolov, N.L., Ivanov, E.A., Kosm Biol
        Aviakosm Med., 1976, 9: 76.
45.   Gordon, S.G., Smith, K., Rabinowitz, J.L., Vagelos, P.R.,
        J. of Lipid Res., 1973, 14: 495.
46.   Labows, J.N., Preti, G., Hoelze, E., Leyden, J., Kligman, A.
        J. Chromatogr. - Biomed. Applic., 1979, 163: 294-299.
47.   Sastry, S., Buck, K., Janak, J., Dressler, M., Preti, G.,
        in Biochemical Applications of Mass Spectrometry ed.
        G. Waller and O.C. Dermer, Wiley, 1980, p. 1086-1123.
48.   Montagna, W., Parakkal., P., Structure and Function of Skin
        Academic Press, 1974.
49.   Hurley, J., Shelley, W., The Human Apocrine Gland in Health
        and Disease, Thomas, Springfield, Illinois, 1960.
50.   Labows, J.N., McGinley, K.J., Leyden, J.J., Webster, G.F.,
        Appl. and Environ. Micro., 1979, 38: 412-415.

51.   Tressel, R., Apetz, M., Arrieta, R., Grunewald, K.,
      in Flavor of Foods and Beverages, ed. G. Charalamborus
      and G. Inglett, Academic Press, 1978, p. 145.
51a. Labows, J.N., McGinley, K.J., Webster, G.F., Leyden, J.J.,
      J. Clin. Microbiol., 1980, accepted for publication.
52.   Cox, C.D., Parker, J., J. of Clin. Micro., 1979, 9: 479-
      484.
53.   Kloek, J., Psychiat. Neurol. Neurochir., 1961, 64: 309.
54.   Kingsbury, A., Brooksbank, B.W.L., Hormone Res., 1978, 9:
      254-270.
55.   Julesz, M., Acta Med., Academy Science Hungar., 1968,
      25: 273.
56.   Brookbank, B.W.L., Experentia, 1970, 26: 1012.
57.   Brookbank, B.W.L., Brown, R., Gustafsson, J.A., Experentia,
      1974, 30: 864.
58.   Gower, D.B., J. Ster. Biochem., 1972, 3: 45.
58a. Bird, S., Gower, D.B., J. Endocrinol, 1980, 85: 8p-9p.
59.   Claus, R., Alsing, W., J. Endocr., 1976, 68: 483.
60.   Labows, J.N., Preti, G., Hoelzle, E., Leyden, J., Kligman, A.
      Steroids, 1979, 3: 249-258.
61.   Amoore, J.E., Pelosi, P., Forrester, J.L., Chemical
      Senses and Flavor, 1977, 2: 401.
62.   Spiteller, G., Pure and Appl. Chem., 1978, 50: 205.
63.   Labows, J.N. Jr., Perfumer and Flavorist, 1979, 4: 12-17.

RECEIVED October 13, 1980.

Structure–Activity Relations in Olfaction

From Single Cell to Behavior—The Comparative Approach

DAVID G. MOULTON

Veterans Administration Medical Center and Department of Physiology,
University of Pennsylvania, Philadephia, PA 19104

Odorants excite receptor cells presumably by interacting
with the hypothetical receptor sites. But to reach these sites,
odorants must first be transported from a point at which their
concentration is known, across the liquid secretions (mucus)
lining the surface nasal of nasal airways, through the mucus/air
phase boundary and possible to the base of the mucociliary blank-
et. The mucus is rich in microproteins, Na^+ ions and pigmented
granules. Within the mucus, odorant molecules may partition between different liquid phases. Thus separate subsets of physiochemical factors govern stages of transport and odorant-receptor
interaction. Consequently, the verbal response - the indicator
used in human psychophysical studies of structure-activity relations - reflects the end product of events whose separate contributions are unknown.
What is needed in interpreting such data, is a means of se-
gregating and manipulating separate phases of the process or of
components of the chemosensory system and assessing their relative influences on the final measured response. To do so we must
turn to animal studies. Thus, in the appropriate preparation, it
is possible to eliminate certain transport factors; to employ an
aqueous rather than a vapor phase to transport odorants to the
olfactory surface; to study separately the response characteristics of subpopulations of receptors differing in their structure-activity relations (including the separate contributions of
the olfactory receptors and the highly chemosensitive endings of
the trigeminal nerve in the nasal mucosa), and to take advantage
of the various anatomical and functional features peculiar to
specific animal groups such as the extreme absolute sensitivity
to certain odorants shown by the dog. The power of the comparative approach to structure-activity relations can be illustrated
with selected examples drawn from electrophysiological studies
in fish, amphibians and mammals; and behavioral studies in mammals. What follows is a selective and not a comprehensive
review of relevant comparative research.

Electrophysiological studies in fish

The most extensive studies on structure-activity relations
in olfaction - apart from those on humans - have been on fish.
This interest relates partly to the commercial importance of this
group. But there are advantages in delivering odorants in the
aqueous phase: sorption onto the fluid secretions (mucus) covering the olfactory surface is likely to be less than occurs with
gaseous odorants and the odorant partition coefficient for water/
mucus will be closer to one than would be the case for air/mucus.
For example, carvone is strongly sorbed anteromedially when
flowed in the vapor phase over the frog's olfactory epithelium
and has a relatively long retention time on this tissue (1,2).
The same compound in the aqueous phase (Ringer's solution) was
flowed over the frog's olfactory olfactory epithelium prior to
washing with tritiated N-ethylmaleimide (NEM - a group specific
protein reagent). Sites protected by carvone from attack by NEM
were subsequently found distributed evenly across the epithelium.
This suggests that sorptive effects do not control odorant distribution in the aqueous phase (3).
     It is true that most of the compounds that normally excite
the olfactory organ in fish differ from those to which air breathing vertebrates are exposed, but there is no evidence that the
basic transduction mechanisms in air and water differ significantly. It is known, for example, that the same odorants delivered
in the aqueous phase, are as effective as when delivered in the
vapor phase as judged by the slow voltage shift recorded when a
macroelectrode tip was positioned in the nasal cavity of a box
turtle during odorant stimulation (4).
     A further advantage in using fish is anatomical. In air-
breathing vertebrates the olfactory chamber extends from the
respiratory airway; in most fish, however, it is a separate organ
divorced from respiratory functions. This feature, and the presence of an aqueous medium, allows us to place a conductivity
electrode at the inlet and one at the outlet of the nasal chamber.
If electrolytes are used as odorants, their arrival and departure
from the chamber can then be measured by conductivity changes.
Since conductivity is proportional to concentration we can specify odorant concentration, within known limits, close to the receptors - something which cannot be done with the intact nasal
chamber in air-breathing vertebrates. It is also possible to
deliver the odorant in a way that closely imitates that in which
it normally arrives (5).
     Among the most effective olfactory stimuli for fish are
amino acids. For example, thresholds of 10^{-9}M have been reported
for L-glutamine in the Conger eel (6) and of 3.2 x 10^{-9}M in the
Atlantic salmon (7) and catfish (8). This sensitivity is presumably related to the widespread distribution of amino acids in

fish skin extracts, which elicit fright and alarm reactions in
other fish of the same species (e.g. 9); in mammalian skin,
which act as a repellant to salmon (10,-13); and in substances
that attract or elicit feeding behavior in fish (14,15). It is
not surprising, therefore, that most structure–activity studies
on fish olfaction have centered on amino acids.

Despite the range of species that have been investigated,
the variety of techniques used and the presence of some species
differences in response (see, for example, 6), there is considerable agreement between workers on the factors that govern neural
response, irrespective of whether activity is recorded at a peripheral or higher level (16-19,6). For example, α-amino acids
elicit the maximum responses, and the most effective member of
a chiral pair is the L-isomer. (Of these, L-glutamine or L-
alanine are the most powerful stimuli for the majority of species
so far tested, but not for all).

An α-amino acid consists of an asymmetrical carbon center
surrounded by four functional groups: (1) α-amino (2) primary-
carboxyl (3) α-hydrogen and (4) a side chain, R:

$$(3)\ H-\overset{\overset{\displaystyle COOH}{|}}{\underset{\underset{\displaystyle R}{|}}{C}}-NH_2\ (1)$$

Response amplitudes can be reduced by substituting other functional groups (-H, -CH$_3$, -OH) for the α-amino group; by methyl-
ation or acetylation of the α-amino moiety; by substitution of
the α-hydrogen; or, in some cases, at least, by replacing the
primary-carboxyl group.

The most effective amino acids are generally those with 5-6
carbon atoms and with linear and uncharged side chains. Amid-
ation greatly increases the effectiveness of aspartic and glut-
amic acid, and sulfur-containing amino acids are also particularly strong excitants. However, Caprio (19) has concluded that,
in general, the S atom may be equivalent to another C atom in the
chain.

The above interpretations of the data do not consider the
alternative implications of a multiple receptor site model for
the odorant-receptor interaction. In such a model the response
elicited by a ligand results from the simultaneous binding of
several groups rather than one. Thus if one group is modified
it may alter the odorant molecule in such a way that it no longer
binds to other sites contributing to the response.

Hara (17) has proposed a model of the amino acid receptor
site consisting of two charged subsites, one cationic and one
anionic, capable of interacting with ionized α-amino and primary-carboxyl groups of amino acid molecules. He assumes that
the L-isomers have more ready access to the receptor and accounts

for this by postulating that the two subsites are arranged around
the third central subsite in such a way that it accommodates the
α-hydrogen atom of an amino acid molecules. Since the fourth
α-amino moiety greatly influences stimulating effectiveness he
proposes that there is a further region which recognizes this
moiety and accounts for discrimating amino-acid quality. Caprio
(19), however, has argued that the binding of the primary carboxyl group may not be primarily ionic.

In the rainbow trout, olfactory bulb neurones seem to discriminate between various chemical stimuli having only slightly
dissimilar molecular structures and conformations. In fact,
several cells, in one study, gave opposite responses to members
of enantiomeric pairs of amino acids: the L-isomer generally
excited while the D-isomer inhibited the cell (18).

There are three problems in particular that complicate in-
terpretation of much of the data on structure-activity relations
in olfaction. First, the different techniques used often yield
data that are not strictly comparable. Recordings from a single
or a few receptors, for example, are more reliable indicators
of the odorant-receptor interaction than are recordings of the
massed action of many neural elements in the olfactory bulb.
Thus discrepancies among results are to be expected. Second,
many workers record without regard to the existence of topo-
graphic differences in the sensitivity of the system to different
odorants. For example, Döving et al (20) showed that bile acids
elicited responses (in the olfactory bulb of chars and graylings)
which differed spatially from those of two amino acids.

A third difficulty is that many workers investigate the response to only one concentration of each odorant. But it is
well known that some odorants can increase neural activity at low
concentrations and suppress it at higher concentrations (21,20,5).
This raises the possibility that the relative stimulating effectiveness of a group of odorants established at one concentration,
is not the same as that existing at another level. The point is
well illustrated in a study by Meredith (22,23,5). The aim was
to establish and analyse response similarities of single bulbar
neurones in the goldfish when stimulated successively by seven
amino acids - each acid being presented in not one, but two different concentrations. The compounds used, their structures and
certain physical properties are shown in Table I.

Response similarity was measured by correlating temporal
patterns of cell firing rate (rather than maximum firing rate -
which was less characteristic of odor type and concentration)
using the Spearman Rank Order Correlation (ρ). (A mean similarity measure for a given stimulus pair was determined by finding
the average firing rate across all units tested. Guttman-Lingoes
nonmetric mulitdimensional scaling procedure was applied to the
matrix representing all pairs of mean similarity measures resulting in the arrangements in Figure(1). In these plots the rank
order of distances between points is the inverse of the rank

Table I.

Physico-chemical Constants and Structures of Amino Acids

Substance	MW	Symbol	Structure	Substance	MW	Symbol	Structure
Glycine	75.1	G	H$-$CHCO$_2$H 　　NH$_2$	Taurine	125.1	T	CH$_2-$CH$_2$SO$_3$H NH$_2$
Alanine	89.1	A	CH$_3-$CHCO$_2$H 　　NH$_2$	Phenyl-alanine	165.2	P	⬡$-$CH$_2-$CHCO$_2$H 　　　　NH$_2$
β-alanine	89.1	B	CH$_2-$CH$_2$CO$_2$H NH$_2$				
Serine	105.1	S	HOCH$_2-$CHCO$_2$H 　　　　NH$_2$	Arginine	174.2	R	NH ‖ C$-$NHCH$_2$CH$_2$CH$_2-$CHCO$_2$H NH$_2$　　　　　　　　NH$_2$

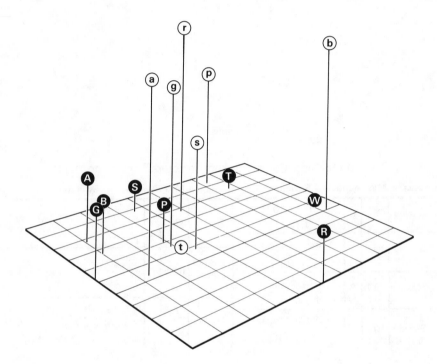

Figure 1.   Multidimensional scaling of responses to a group of amino-acids.

The responses ultimately are derived from the temporal firing patterns of single cells in the goldfish olfactory bulb. Symbols for acids presented at 10^{-2}M are shown in large case while symbols for acids presented at 10^{-4}M are shown in small case. (i.e. higher concentrations are low in the space while lower concentrations are high in the space). G,A,S and to a lesser extent P, form a related group whose distances from one another (response similarities) are relatively constant across concentrations. The distances between R,B and the other acids are markedly altered by changing concentration. (W, the tap water in which the fish were kept, is included as a control substance, but is likely to contain amino acids originating from the fish themselves). For key to symbols see Table I. Data are from Meredith (23), and the multidimensional scaling analysis was performed by S. S. Schiffman who used a nonmetric method that involves no assumption about the underlying dimensions used (see text).

order of similarity of response to stimulus pairs ($A_v e_s$ values).
Acids eliciting similar responses are thus, on average, closer
than those giving dissimilar responses which are at the opposite
ends of the space. Clearly the distances between alanine (A) and
β-arginine (B) are markedly altered upon changing concentration.
The conclusion is that response similarities measured at one concentration do not necessarily predict those existing at other
concentrations. A similar conclusion was reached on the basis
of animal psychophysical studies (24).

Fig. 1 illustrates a further point: distances between some
compounds are consistent at both concentrations tested, for compounds which are either proximate to one another (alanine, glycine and serine) or distant from one another (e.g. alanine and
taurine). As Meredith (23) points out, the persistence of response similarities is explicable in terms of the molecular
structure of the three amino acids which differ only in the substitution of $-CH_3$ and $-CH_2OH$ for $-H$ on the α-carbon of the glycine. Consequently they may activate the same receptor sites.
The amino acid, phenylalanine, with aromatic ring, although
close at 10^{-2} M, falls slightly separate from the aliphatic amino
acids. The complete separation of taurine from the other compounds is most probably related to the sulfur atom. This arrangement is similar to one derived for tastes of amino-acids
(25,26,27). For example, serine, alanine and glycine are sweet
and cluster together in space while taurine is bitter and falls
out in space.

Single unit studies in amphibians

Because of the central importance of human olfaction to
many investigators interested in structure-activity relations,
mammals are often the animals of choice in electrophysiological
studies. Unfortunately, the bony turbinates, which support the
olfactory receptor sheet in mammals, are often elaborately con-
voluted. This greatly complicates the problem of delivering the
stimulus to the receptors in a controlled fashion. In contrast,
some amphibians have an unfolded olfactory epithelium and odorants are easily directed to any point on the surface. In particular, the tiger salamander has a relatively flat receptor
surface and is increasingly becoming the animal of choice in in-
vestigations of receptor properties. However, many workers continue to use the frog despite a domed region in its epithelium.

One study analyzed the responses of single olfactory receptors in the frog to a group of 20 odorants. The odorants
tended to fall into four groups: (i) An aromatic group including
benzene, anisole, bromobenzene and dichlorobenzene. (Responses
to thiophenol showed some relation to this group). (ii) Camphor
and cincole (iii) Cyclohexanol, cyclohexanone and tert-butanol
(iv) A fatty acid group consisting of butyric, valeric and isovaleric acids. Thiophene fell outside these groups (28,29). One

surprizing feature was the relatively poor effectiveness shown
by sulfur compounds. In fact thiophene, butanethiol-1, and
diethyl sulfide failed to elicit any measureable response in
most of the frogs. The authors suggest that frog receptors
lack sites for S or S-H groups. The relatively greater stimulating effectiveness of thiophenol may stem from its benzene nucleus
rather than any contribution from its S-H group (28).

There are differences among frog receptor cells in their
ability to discriminate among sterically related odorants: One
cell was excited by p-tolyurea but not by o- or m-tolyurea.
Other receptor cells did not discriminate among these isomers
(30). In general, most workers report that although some receptor cells do not respond to any odorant (31), many respond to the
majority of odorants tested. But although the receptor cell as
a whole may have low odor specificity this does not eliminate
the possibility that it may possess several or many different
types of receptor site each of which might show a high degree of
specificity for a given odorant. It could be argued, then, that
the ultimate target in the study of structure-activity relations
is the receptor site - odorant interaction. Is there any method
that might give some insight into the numbers, kinds and properties of site types? One promising approach does exist. (It
exploits the phenomenon of cross-adaptation to odorants. (Olfactory cross-adaptation is the decline in response magnitude
to an odorant that occurs as a result of prolonged exposure to
another odorant). The idea is that if a receptor contains at
least two types of site sensitive to odorants A and B respectively, it should be possible to adapt out those sites sensitive to
A. If the second odorant B is now delivered the response of the
cell will depend on whether B occupies the same or different
sites. Response amplitude to B will be reduced (relative to the
control response to B) if the sites are the same, but remain
unchanged if the sites are different.

This was the approach used by Baylin and Moulton (32) in
studying the properties of single epithelial cells in the tiger
salamander. They tested seven odorants in 4 pairs in which the
odors of members of a pair are similar - at least, to humans.
(The pairs were methyl butyrate and ethyl butyrate, butanol and
propanol; benzaldehyde and nitrobenzene; and benzaldehyde and
acetophenone. An odorant pulse lasted 5 sec. Each odorant was
delivered as a single pulse as two successive pulses, and paired
with a pulse of the second odorant as follows: A/A,A/A,B/B/B,B/B,
A. All pairs of pulses were separated by t secs, where t=0-10
secs and was fixed in any sequence but varied between experiments.
The A,A and B,B pairings gave measures of self-adaptation while
the A,B and B,A pairings gave measures of cross-adaptation).

A given cell showed either self- or cross-adaptation to both
or to either memebers of a pair of odorants. But the most striking finding was that all cross-adaptation was nonreciprocal.
Thus, in some cells A adapted B (but not vice versa), while in

others the reverse occurred. Cross-adaptation could occur independently of self-adaptation and in some receptors neither occurred. In addition, receptors were found which responded to either
butanol or propanol but not both. In fact, the data suggested
that, with one possible exception, receptive sites existed which
responded to each of the seven odorants tested.

Baylin and Moulton suggest that the simplest model for
cross-adaptation that could explain those findings assumes that
sites exist that respond to A alone, to B alone and to both A
and B. Thus, if B sites were absent from a cell it would respond
to A and B separately, but while A would adapt B, B could not
adapt A.

Spatial patterning of response to odorants

The rod and cone receptors of the eye are spatially segregated on the retinal surface. If the olfactory surface were or-
ganized according to similar principles, it would greatly facil-
itate the analysis of structure-activity relations. No such
sharp separation has yet emerged. Yet the olfactory receptors
do show some spatial segregation according to their odor specificities, even though it is not absolute. In fact, odorants tend
to fall into three broad categories depending on the spatial
gradient of excitation that they elicit in the olfactory epithelium of the tiger salamander. Most are more effective anteriorly than posteriorly; some show the reverse pattern and a few
cannot be classified into either category but seem to stimulate
more uniformly. In the case of butanol (anterior stimulant) and
limonene (posterior stimulant) the average composite difference
in sensitivity exceeds one order of magnitude (33). The classification, structures and physical properties of odorants so far
tested are summarized in Figure 2. From this it is clear that
posterior stimulants differ from all other odorants in combining
both insolubility in water with solubility or complete miscibil-
ity in alcohol. A full structure-activity analysis is not war-
ranted until the regional distribution of stimulating effectiveness for more odorants have been measured. Nevertheless, there
is a suggestion here that lipophilicity may be a significant
factor controlling structure-activity relations in some types of
odorant-receptor interaction.

These categories provide only an initial sorting of odorants based on a comparison of response magnitudes recorded electrophysiologically from two micropipette positions (one anterior
and the other posterior). When 30 epithelial positions are
sampled in this way, a map can be generated for each odorant.
Such maps show a further type of more specific spatial patterning superimposed on the general anterior-posterior organization.
This takes the form of small zones of greatly heightened sensitivity. Among odorants so far tested, the shape and position of
these regions is specific for one or two odorants. For example,

Anterior stimulants

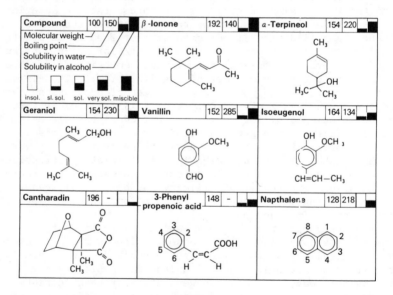

Figure 2. Structures and selected properties of odorants used to stimulate olfactory epithelium of tiger salamander. Odorants are classified according to their relative effectiveness in stimulating different epithelial regions (33).

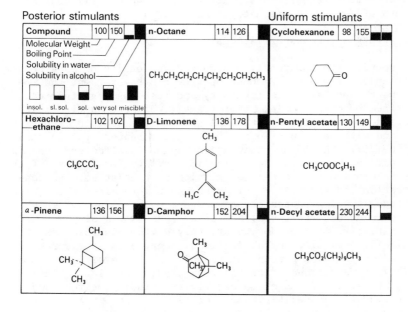

Figure 2.   Continued

pentyl acetate generally stimulates maximally along a distinct
ridge extending anteriorly for about 2mm from the d-limonene
sensitive zone located posteriorly. Foci for peak sensitivity
to eugenol and iso-eugenol are also clearly segregated (34).

     In this study odorants were delivered by way of a micro-
pipette positioned less than a mm from the olfactory surface (a
modification of a technique described by Kauer and Moulton (35)).
This eliminates any possibility that the differences reported resulted from a differential sorption of odorant molecules by the
mucus. In the behaving animal, however, odorants flow over the
mucus anteriorly to posteriorly. Because different odorants have
different mean relative retention times on the olfactory mucosa
(1 ,36) they may create different gradients of excitation across
it (except for those with relatively short mean retention times).
To what extent such factors contribute to the overall spatial
pattern of excitation is not yet clear (37).

Concentration-response relations

     Despite the apparently widespread conformity of many sensory
functions to the Weber-Fechner or Steven's Power Laws, the relations between stimulus intensity and response magnitude can
sometimes be more complex. For example, discontinuities in this
relation are associated with dual somesthetic receptor functions
(38) and with dual functions of a single receptor type in the
retina (39). Should such deviations occur in olfactory functions,
they may not have been identified in many studies because of the
very assumption that a simple relation must exist between concentration and response. This assumption determines the concentrations at which response measures are made - a number which
may be inadequate to reveal any deviations from a simple relation.
Alternatively, if they do appear, they may be dismissed as statistically insignificant aberrations. Yet, even if a curve reflected only ligand binding it is unlikely to be simple. In a
variety of neural and other tissues, binding curves are influ-
enced by various froms of cooperativity (binding, effect and
intermolecualr cooperativity). For example, in binding cooperativity the presence of ligand molecules already bound can alter
the affinity of the receptor for additional ligand binding (40).
But in fact, further complexity may be imposed by events pre-
ceding and succeeding odorant binding. The most significant of
these are transport factors and nonlinear transform functions
within the central nervous system (in the case of measures taken
at bulbar or higher levels). It has also been suggested that
enzymes, which are probably present in the mucus, may degrade
odorant molecules diffusing towards binding sites (Nicollini
41).   Thus as Getchell and Getchell (42) have noted, pentyl
acetate may degrade to pentanol and acetic acid. These events
could further distort the concentration-response curve.

     In view of these factors it is not surprizing that single

cells in the goldfish olfactory bulb yield curves with a variety
of shapes. Some, for example, show monotonic response functions
while others show initially increasing and then decreasing firing
rates as concentration is increased (Fig. 3). Some non-monotonic
response functions were also recorded in frog and salamander olfactory receptors (43,44). The average of concentration response-
curves showing a variety of forms, such as those in Fig. 3, could
be a relatively complex function. But whatever the reason, for
some odorants at least, curves derived from large populations of
receptors do show marked notches. They appear in data generated
both electrophysiologically and psychophysically (Fig. 4). In
the case of the psychophysical curve for α-ionone seen in data
from dogs, the notch is highly significant statistically and
divides the curve into a slowly descending upper limb, best fitted
by a parabolic function, and a rapidly descending lower limb, best
fitted by a cubic function (Fig. 5). In an homologous series of
aliphatic acetates the position of this notch on the curve ascends
with increasing chain length, and it has been suggested that the
notch may reflect the independent contributions of two types of
receptors - the response of one, controlling the lower limb of
the curve, and that of the other, controlling the form of the
upper limb (24). An alternative explanation, however, is that
the affinity of a single type of site for the odorant changes as
a critical concentration is reached.

The form of the concentration-response curve offers a potential approach to grouping odorants, and Mathews (48) has made a
promising start in this direction. He recorded the averaged activity from bundles of receptor nerve fibers in the rat. The
seven odorants he tested fell into three groups according to the
slope and form of the curves that they elicited. Members of the
first group were n-pentyl acetate and two compounds with a pepper-
minity odor: menthone and 2-sec butyl hexanone. Their curves
showed clear notches and accelerated negatively towards their
asymptotes. The second group contained linalool and dimethyl
benzyl carbonyl acetate - compounds with a floral odor and posi-
tively accelerating curves. In the third group were camphor and
iso-borneol, both with a camphoraceous odor and a curve consisting of a lower negatively accelerating limb and a linear upper
limb.

Further evidence that the slope of the concentration-response
curve may be related in a predictable way to the physiochemical
properties of the odorant molecule comes from a study of the relative detectability of members of an homologous series of aliphatic acetates (49). Probit regression lines were first derived
from the concentration-response data for each member of the
series. The slopes of these lines, when plotted against log
carbon chain length, yielded an approximately linear relation.
One consequence of the relation is that each of several probit
regression lines intercepts one or more other lines. Thus the
relative effectiveness of these compounds depends on the perfor-

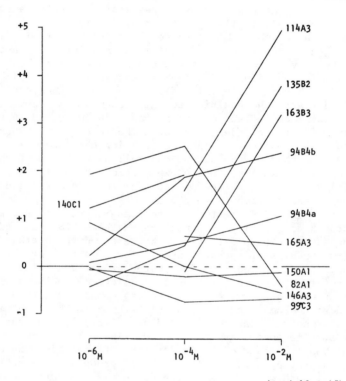

Journal of General Physiology

Figure 3.   Concentration–response relations of 11 units to glycine.   Magnitude of response, measured as the normalized average firing rate for the rise and plateau phases of the stimulus, is plotted on the ordinate (5).

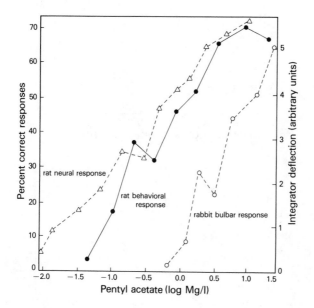

*Figure 4.   Comparison of concentration–response functions for amyl acetate de-
rived from psychophysical and electrophysiological measures of response. The
partially overlapping curves are for the rat—one was generated by rats performing
on an odor choice apparatus (24) while the other reflects the massed responses of
receptor nerve bundles (45). The remaining curve is the averaged multiunit activity
of the rabbit olfactory bulb (46).*

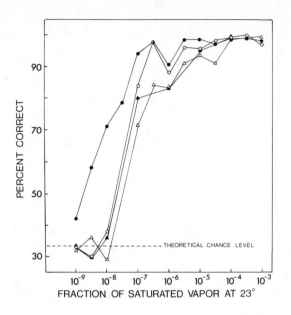

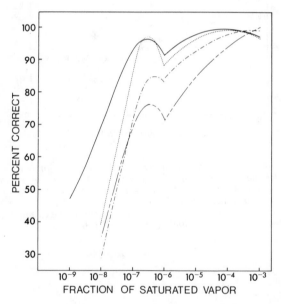

Journal of Comparative Physiology

Figure 5. (a) Performance of 4 dogs in detecting α-ionone in the vapor phase (dog no. 1 (●—●); 2 (○—○); 3 (▲—▲); 4 (△—△)); (b) least-squares curve fits to the data shown in (a), assuming the response function is the sum of two distinct processes (dog no. 1 (——); 2 (– – –); 3 (– – – – –); 4 (— · — ·)) (47).

mance level that is chosen as a basis for comparison.   For example, if a 50% correct response score is taken as the criterion (chance being 50% correct), the relation between performance and response is linear.   If, however, an 85% correct criterion is chosen a partially curvilinear relation emerges (49).   The dependence of response similarities (determined electrophysiologically) on concentration was discussed above in relation to a group of amino acids.   Thus the physicochemical properties of an odorant that control its relative stimulating effectiveness at one concentration are not necessarily those controlling effectiveness at another concentration.

Acknowledgments

We thank Dr. Michael Meredith for permission to include figures and data from a presentation and for critically reading the manuscript.   We also thank Dr. Susan Schiffman for permission to include the results of a multidimensional analysis of the data and for helpful discussion and comments on the manuscript.   Part of the work described here was supported by NIH grant No. 5 R01 NS 10617-04 and AFOSR grant No. 77-3162.

Literature Cited

1. Mozell, M.M. Evidence for a chromatographic model of olfaction. J. of Gen. Physiol., 1970, 46-63.
2. Mozell, M.M. and Jagodowicz, M. Chromatographic separation of odorants by the nose: retention times measured across in vivo olfactory mucosa. Science, 1973, 1247-1249.
3. Nathan, M.H. The localization of sulfhydryl groups and maping of specific olfactory receptor sites in the frog olfactory epithelium. Ph.D. Thesis, The University of Michigan, 1979.
4. Tucker, D. and Shibuya, S. A physiologic and pharmacologic study of olfactory receptors. Cold Spring Harbor Symposium of Quantitative Biology, 1965, 30, 207-215.
5. Meredith, M. and Moulton, D.G. Patterned response to odor in single neurones of goldfish olfactory bulb: influence of odor quality and other stimulus parameters. J. of Gen. Physiol., 1978, 71, 615-643.
6. Goh, Y., Tamura, T. and Kobayashi, H. Olfactory responses to amino acids in marine teleosts. Comp. Biochem. Physiol., 1979, 62A, 863-868.
7. Sutterlin, A.M. and Sutterlin, N. Electrical responses of the olfactory epithelium of atlantic Salmon (salmo salar). J. Fish. Res. Bd. Canada, 1971, 28, 565-572.
8. Suzuki, N. and Tucker, D. Amino acids as olfactory stimuli in fresh water catfish Ictalurus catus (Linn.). Comp. Biochem. Physiol., 1971, 40A, 399-404.
9. Tucker, D. and Suzuki, N. Olfactory responses to Schreckstoff of catfish. In: Schneider, D., Ed. "Olfaction and Taste IV." Wissenschaftliche Verlagsgesellschaft, Stuttgart, 1972, 121-134.
10. Idler, D.R., Fagerlund, U.H. and Mayoh, H. Olfactory perception in migrating salmon I. L-serine, a salmon repellent in mammalian skin. J. Gen. Physiol., 1956, 39, 889-892.
11. Idler, D.R., McBride, J.R., Jonas, R.E.E. and Tomlinson, N. Olfactory perception in migrating salmon II. Studies on a laboratory bioassay for homestream water and mammalian repellent. Can. J. Biochem. Physiol., 1961, 39, 1575-1584.
12. Hamilton, P.B. Amino acids on hands. Nature, 1965, 205, 284-285.
13. Oro, J. and Skewes, H.B. Free amino acids on human fingers: the question of contamination in microanalysis. Nature, 1965, 207, 1042-1045.
14. Pawson, M.G. Analysis of a natural attractant for whiting Merlangius merlangus L. using a behavioral bioassay. Comp. Biochem. Physiol., 1977, 56A, 129-135.
15. Konosu, S., Fusetani, N., Nose, T. and Hashimoto, Y. Attractants for eels in the extracts of short-necked clam - II. Survey of constitutents eliciting feeding behavior by fractionation of the extracts. Bull. Jap. Soc. Sci. Fish., 1968, 34, 84-87.

16. Hara, T.J. Olfaction in fish. Progress in Neurobiol., 1975, 5, 271-335.

17. Hara, T.J. Further studies on the structure-activity relationships of amino acids in fish olfaction. Comp. Biochem. Physiol., 1977, 56A, 559-565.

18. MacLeod, N.K. Spontaneous activitiy of single neurons in the olfactory bulb of the rainbow trout (Salmo gairdneri) and its modulation by olfactory stimulation with amino acids. Exp. Brain Res., 1976, 25, 267-278.

19. Caprio, J. Olfaction and taste in the channel catfish: an electrophysiological study of the responses to amino acids and derivatives. J. Comp. Physiol. A, 1978, 123, 357-371.

20. Doving, K.B., Selst, R. and Thommsen, G. Olfactory sensitivity to bile acids in salmonid fishes. Acta. Physiol. Scand., 1980, 108, 123-131.

21. Kauer, J.S. Response patterns of amphibian bulb neurones to odor stimulation. J. Physiol. (London), 1974, 243, 695-712.

22. Meredith, M. Olfactory coding: single unit response to amino acids in goldfish olfactory bulb. Ph.D. Thesis, University of Pennsylvania, 1974.

23. Meredith, M. Multidimensional scaling of olfactory bulb unit data. Presentation at 1st Meeting of Association for Chemoreception, Sarasota, Florida, 1979.

24. Moulton, D.G. Studies in olfactory acuity 3. Relative detectability of n-aliphatic acetates by the rat. Quart. J. Exp. Psychol., 1960, 12, 203-213.

25. Schiffman, S.S. and Dachis, C. Taste of nutrients: amino acids, vitamins and fatty acids. Percept. and Psychophys., 1975, 17, 140-146.

26. Schiffman, S.S. and Engelhard, H.H. Taste of dipeptides. Physiol. and Beh., 1976, 17, 523-535.

27. Yoshida, M. Similarity of different kinds of taste near threshold concentration. Jap. J. Psychol., 1963, 34, 25-35.

28. Revial, M.F., Duchamp, A. and Holley, A. Odour discrimination by frog olfactory receptors: a second study. Chem. Senses and Flavour, 1978, 3, 7-21.

29. Revial, M.F., Duchamp, A., Holley, A. and MacLeod, P. Frog olfaction: odour groups, acceptor distribution and receptor categories. Chem. Senses and Flavour, 1978, 3, 23-33.

30. Getchell, T.V. Unitary responses in frog olfactory epithelium to sterically related molecules at low concentrations. J. Gen. Physiol., 1974, 64, 241-261.

31. Juge, A., Holley, A. and Delaleau, J.C. Olfactory receptor cell activity under electrical polarization of the nasal mucosa in the frog. J. Physiol.(Paris), 1979, 75, 919-927.

32. Baylin, F. and Moulton, D.G. Adaptation and cross-adaptation to odor stimulation of olfactory receptors in the tiger salamander. J. of Gen. Physiol., 1979, 74, 37-55.

33. Kubie, J.L. and Moulton, D.G. Regional patterning of re-

sponse to odors in the salamander olfactory mucosa. Neuroscience Absts., 1979, 5, 129.

34. Mackay-Sim, A. and Moulton, D.G. Odorant specific maps of
relative sensitivity inherent in the salamander olfactory
epithelium. Neuroscience Absts. 6, 1980 (in press).

35. Kauer, J.S. and Moulton, D.G. Responses of olfactory bulb
neurones to odour stimulation of small nasal areas in the
salamander. J. Physiol., 1974, 243, 717-737.

36. Hornung, D.E. and Mozell, M.M. Factors influencing the differential sorption of odorant molecules across the olfactory
mucosa. J. Gen. Physiol., 1977, 69, 343-361.

37. Moulton, D.G. Spatial patterning of response to odors in the
peripheral olfactory system. Physiol. Rev., 1976, 56, 578-
593.

38. Talbot, W.H., Darian-Smith, I., Kornhuber, H.H. and Mount-
castle, V.B. The sense of flutter-vibration: comparison of
human capacity with response patterns of mechanoreceptive
afferents from monkey hand. J. Neurophysiol., 1968, 31, 301-
334.

39. Green, D.G. and Siegel, I.M. Double-branched flicker fusion
curves from the cone-rod skate retina. Science, 1975, 188,
1120-1122.

40. Stevens, C.F. Biophysical analysis of the function of receptors. Ann. Rev. Physiol., 1980, 42, 643-652.

41. Niccolini, P. Lo stimolo olfattorio et la sua recezione.
Arch. Ital. Sci. Fharmacol., 1954, 4, 109-172.

42. Getchell, T.V. and Getchell, M.L. Early events in vertebrate olfaction. Chem. Senses and Flavour, 1977, 2, 313-326.

43. Gesteland, R.C. Receptor membrane functions in chemorecep-
tion: implications from single unit studies. In: Benz, G.,
Ed., "Structure-activity Relationships in Chemoreception."
Information Retrieval, London, 1976, pp. 161-168.

44. Baylin, F. Temporal patterns and selectivity in the unitary
responses of olfactory receptors in the tiger salamander to
odor stimulation. J. Gen. Physiol., 1979, 74, 17-36.

45. Mathews, D.F. Response patterns of single neurones in the
tortoise olfactory epithelium and bulb. J. Gen. Physiol.,
1972, 60, 166-180.

46. Mozell, M.M. Electrophysiology of the olfactory bulb. J.
Neurophysiol., 1958, 21, 183-196.

47. Moulton, D.G. and Marshall, D.A. The performance of dogs in
detecting α-ionone in the vapor phase. J. Comp. Physiol.,
1976, 110, 287-306.

48. Mathews, D.F. Rat olfactory nerve responses to odors. Chem.
Senses and Flavour, 1974, 1, 69-76.

49. Moulton, D.G. and Eayrs, J.T. Studies in olfactory acuity
2. Relative detectability of n-aliphatic alcohols by the
rat. Quart. J. Exp. Psychol., 1960, 12, 99-109.

Received October 13, 1980.

INDEX

INDEX